Health Policy and High-Tech Industrial Development

Learning from Innovation in the Health Industry

Edited by

Marco R. Di Tommaso

Professor of Industrial and Development Policy, University of Ferrara, Italy and Deputy Director, L'institute (Institute for Industrial Development Policy)

Stuart O. Schweitzer

Professor of Health Services, School of Public Health and Associate Director, Program in Pharmaceutical Economics and Policy, University of California, Los Angeles, USA

IN ASSOCIATION WITH L'INSTITUTE (INSTITUTE FOR INDUSTRIAL DEVELOPMENT POLICY), UNIVERSITIES OF BIRMINGHAM (UK), FERRARA (ITALY) AND WISCONSIN-MILWAUKEE (USA)

Edward Elgar

Cheltenham, UK • Northampton, MA, USA

Published by
Edward Elgar Publishing Limited
Glensanda House
Montpellier Parade
Cheltenham
Glos GL50 1UA
UK

Edward Elgar Publishing, Inc.
136 West Street
Suite 202
Northampton
Massachusetts 01060
USA

A catalogue record for this book
is available from the British Library

ISBN 1 84376 757 0

Printed and bound in Great Britain by MPG Books Ltd, Bodmin, Cornwall

Contents

List of figures vii
List of tables and box ix
List of contributors xi
Acknowledgements xiii

PART ONE: AN OVERVIEW

Introduction: why apply industrial policy to the health industry? 3
Marco R. Di Tommaso and Stuart O. Schweitzer

1. The Health Industry Model: new roles for the health industry 17
Stuart O. Schweitzer and Marco R. Di Tommaso

PART TWO: THE MACRO VIEW

2. Healthy governance: economic policy and the Health Industry Model 45
J. Robert Branston, Lauretta Rubini, Roger Sugden and James R. Wilson
3. Control, competition and co-operation in European health systems: points of contact between health policy and industrial policy 59
Giovanna Vicarelli
4. A hedonic model of pricing of innovative pharmaceuticals 77
William S. Comanor, Stuart O. Schweitzer and Tanja Carter

PART THREE: THE MICRO VIEW

5. Recent developments in universities regarding intellectual capital and intellectual property 103
Emidia Vagnoni, James Guthrie and Peter Steane
6. Intangible assets in the European health industry: the case of the pharmaceutical sector 125
Patrizio Bianchi and Sandrine Labory

7. Benchmarking hospital costs in the UK: increasing efficiency and driving innovation in a health care industry? 150
Sue Llewellyn and Deryl Northcott

PART FOUR: THE INTERMEDIATE VIEW

8. The geography of intangibles: the case of the health industry 171
Marco R. Di Tommaso, Daniele Paci and Stuart O. Schweitzer

9. Clustering in the biotechnology industry 206
Stuart O. Schweitzer, Judith Connell and Fredrick P. Schoenberg

10. Spillovers of university–high-tech industry alliances 220
Werner Z. Hirsch

11. Multinational enterprises and high-tech clusters in the health industry: some preliminary results in Italy 233
Marco Bellandi and Nicoletta Tessieri

12. High-technology clusters in France: two unusual models – an empiric study 258
Grégory Katz-Bénichou and Gérard Viens

Index 279

Figures

0.1	The industrial policy debate	5
0.2	The 'health industry' model	8
1.1	Public share of health expenditures in the G-7 countries, 1960–2000	24
1.2	The 'health industry' model	26
1.3	Ownership of hospital beds in the USA, 1980–2000	28
1.4	The rising number of the elderly (aged 65+) in the G-7, 1986–96	35
1.5	The rising number of very elderly (aged 80+) in the G-7, 1986–96	35
3.1	Graphic representation of the three traditional forms of social regulation	72
3.2	Graphic representation of new and traditional forms of social regulation	73
5.1	Groupings within the organizational knowledge framework	107
6.1	New chemical or biological entities, 1991–2000	137
8.1	Intangible resources	175
8.2	The new corporate asset base	175
8.3	Intangibles and the production process	176
8.4	Knowledge capital/book value in selected companies, 30 Sept. 2000	178
8.5	Knowledge capital/book value comparison among industries, 30 Sept. 2000	180
8.6	Market value/book value comparison among industries, 30 Sept. 2000	180
8.7	Perspectives on intangibles	188
10.1	Three impact stages of university–high-tech industry research alliances	223
10.2	Three stages of California's research alliances' economic impact on California	230

Tables and box

TABLES

1.1	Share of GDP allocated to health in the G-7 countries	18
1.2	Per capita health expenditures in the G-7 countries	19
1.3	Health services employment and share of total employment in the G-7 countries, selected years	20
1.4	Employment in selected sectors in the G-7 countries, 1996	20
1.5	Selected health service inputs and capabilities among the G-7 countries, 2000	23
1.6	Public share of health expenditures among the G-7 countries	23
1.7	Proportion of national health expenditures spent on selected health care providers: inpatient hospital services and ambulatory physician services in the G-7 countries	27
1.8	Proportion of spending on pharmaceuticals in the total health expenditures in G-7 countries	31
1.9	Spending on pharmaceutical R&D in selected countries, 1998	32
1.10	Projected proportion of older people in the total population, 1996–2050	34
4.1	Antidepressants: factors determining 1997 prices paid by HMOs and pharmacies	88
4.2	Calcium channel blockers: factors determining 1997 prices paid by HMOs and pharmacies	90
4.3	Antimicrobials: factors determining 1997 prices paid by HMOs and pharmacies	92
4.4	Antihistamines: factors determining 1997 prices paid by HMOs and pharmacies	94
4.5	Glaucoma opthalmic solutions: factors determining 1997 prices paid by HMOs and pharmacies	96
5.1	Share of total business enterprise funds and government funds in the university R&D sector	115
5.2	University of Ferrara, share of business enterprise funds and government funds in the university R&D areas	121
6.1	A taxonomy of the determinants of a firm's value	133
6.2	Biotech companies: Europe versus USA	137

6.3 Patent applications at the European Patent Office 137
6.4 Activity in genetic engineering, by type of institution 138
6.5 The changing structure of company costs in the pharmaceutical industry, 1973–89 139
6A.1 Lev's estimation of knowledge capital 147
6A.2 Top ranking companies in the health industry 148
7.1 The 'ladder of success' for NHS hospitals 156
7.2 Key national reference cost statistics 157
8.1 Capital stock and capital–output ratio in the USA, 1929–90 173
8.2 Knowledge capital estimations for selected companies, 30 Sept. 2000 177
8.3 Knowledge capital estimations, comparisons among industries, 30 Sept. 2000 179
8.4 R&D indicators for pharmaceutical and biotech companies 181
8.5 The changing structure of company costs in the pharmaceutical industry, 1973–89 181
9.1 Results of the Poisson regressions predicting the likelihood of a biotech firm in a zip code 214
9.2 Variance–covariance matrix 215
9.3 Distance between simulated and actual firms 215
10.1 Economic impact of UC San Diego's corporate research support (and royalties plus fees) in 1998 227
10.2 Economic impact of University of California's corporate research support (and royalties plus fees) in 1998 228
10.3 Economic impact of California's research universities' corporate research support (and royalties plus fees) in 1998 229
11.1 HIBs by regions, provinces and local systems 239
11.2 Types of local systems involved by the presence of HIBs 241
11.3 Headquarters and manufacturing plants belonging to HIBs, by types of local systems 242
11.4 Local systems belonging to 1LS, by types, with pharmaceutical and/or biomedical specialization indexes higher than 2.0 242
12.1 Comparison between biotechnology industries in 2002 262
12.2 Development of biotechnologies in Europe: comparative analysis, 2001 262
12.3 The burden of taxation in France 263
12.4 Financial aspects, 2000 269
12.5 Total investment in the Genopole cluster, 2000 269

BOX

11.1 Sources, legends and terminology 257

Contributors

Marco Bellandi, Università degli Studi di Firenze, Italy

Patrizio Bianchi, University of Ferrara, Italy and L'institute

J. Robert Branston, University of Bath, UK and L'institute

Tanja Carter, Pepperdine University, Malibu, California, USA

William S. Comanor, University of California, Los Angeles and Santa Barbara, USA

Judith Connell, University of California, Los Angeles, USA

Marco R. Di Tommaso, University of Ferrara, Italy and L'institute

James Guthrie, University of Sydney, School of Business, Sydney, Australia

Werner Z. Hirsch, University of California, Los Angeles, USA

Grégory Katz-Bénichou, Co-holder of the Aventis Chair in Ethics and Biotechnology, ESSEC Business School, Paris, France

Sandrine Labory, University of Ferrara, Italy

Sue Llewellyn, The University of Edinburgh, UK

Deryl Northcott, The Auckland University of Technology, New Zealand

Daniele Paci, University of Ferrara, Italy

Lauretta Rubini, University of Birmingham, UK, University of Ferrara, Italy and L'institute

Frederick P. Schoenberg, University of California, Los Angeles, USA

Stuart O. Schweitzer, University of California, Los Angeles, USA

Peter Steane, Macquarie Graduate School of Management, Sydney, Australia

Roger Sugden, University of Birmingham, UK and L'institute

Nicoletta Tessieri, Università degli Studi di Firenze, Italy

Emidia Vagnoni, University of Ferrara, Italy

Giovanna Vicarelli, Università Politecnica delle Marche, Ancona, Italy

Gérard Viens, ESSEC Business School, Paris, France

James R. Wilson, University of Birmingham, UK and L'institute

Acknowledgements

This book is a product of the *Health Industry Policy Project*, a three-year research effort conducted by members of the Faculty of Economics of the University of Ferrara (Italy) and by L'institute, the Institute for Industrial Development Policy. The research project has involved academics, entrepreneurs and policy makers and has benefited from the contributions of an international network of scholars working in Europe, the USA, Australia and China.

The project has been supported by the Fondazione Aventis and Farmindustria under the auspices of the Italian Ministry of Education, Research and Universities, the Italian Ministry of Industry, the Region of Emilia Romagna and the City of Ferrara.

We are indebted to many people for contributing to this effort. First, we would like to express our gratitude to Professor Patrizio Bianchi of the University of Ferrara and L'institute for his unfailing encouragement and creative suggestions for conducting our project. We also wish to thank Professor Roger Sugden of the University of Birmingham and L'institute for his stimulating participation in our Project's several Forums and for his valuable suggestions during the development of this book.

A particular expression of our gratitude goes to Pierluigi Vigo, President of the Fondazione Aventis, and Gianfranco Leoni, the former President of Farmindustria, for their inspiration and support at each stage of our Project. Their warmth and friendship has been invaluable to us.

Francine O'Sullivan of Edward Elgar Publishing has given us considerable guidance in the preparation of the book, and Joy Lorenzana has contributed excellent editorial services.

Finally, we wish to thank Dr Lauretta Rubini of the University of Ferrara and L'institute for her intellectual and organizational input throughout the Project and for her assistance with the preparation of this book. Neither the Project nor this book would have been possible without her wise counsel, always cheerfully given, even during the most stressful periods.

PART ONE

An Overview

Introduction: why apply industrial policy to the health industry?

Marco R. Di Tommaso and Stuart O. Schweitzer*

INTRODUCTION

An important goal of most industrialized countries is to promote its high-technology sectors, such as aerospace, pharmaceuticals, biotech, and telecommunication. These countries fear being left behind in the race for supremacy in each of these fields. An important issue facing any country desiring to compete is the appropriate role of government in these efforts. Many countries have a long tradition of promoting favoured industries in Europe and Asia, such as Italy, France, Germany, Japan and Korea. The USA has less of an explicit attitude toward this policy. Nonetheless, even the USA has engaged in strong government-supported industrial assistance programmes in areas including aerospace, aircraft, computer chips and flat display screens. Approaches are often less direct than those in Europe and Asia, but they are no less strong.

WHY SHOULD HIGH-TECHNOLOGY INDUSTRIES BE PROMOTED?

The issue of industrial promotion is one of the most debated in our recent history (Deane, 1989). However, it should be emphasized that this issue has taken on special significance in recent years because of the growth of high-technology, knowledge-based industries, and new answers are sought to develop, foster or protect these new industries.[1]

Dominant economic theory has reached the conclusion that, with few exceptions, government intervention is likely to lead to less, rather than more, efficiency: in this context, according to mainstream Economics, there is no need for industrial promotion policy. Nonetheless, if one looks at the last two centuries of our industrial development history, industrial policy has always existed and it continues to exist in almost every country (Gerosky, 1989; Chang, 1994; Cowling, 1999).[2]

Different perspectives, or points of view, exist pertaining to the appropriateness of industrial policy, and it is difficult to develop an exhaustive taxonomy of rationales. However, in synthesis, we may argue that it has typically been justified on two grounds, as we show in Figure 1.

The first is economic efficiency. This argument, rooted in the dominant economic theory, is that markets will sometimes fail, for various reasons, and intervention may be needed to restore efficient allocation of resources (Musgrave and Musgrave, 1984; Stiglitz, 1988). Basic textbooks of economic theory justify government intervention primarily to address inefficiencies caused by *market failures*. These market failures result from the existence of public goods, externalities, declining long-run average costs and insufficient competition. In this context, industrial policy may be called upon to correct these market failures. In the case of high-tech sectors, these interventions are often related to research, innovation, infrastructure, local development, antitrust policy and regulation. For example, market forces in some circumstances may not guarantee enough innovative capacity in an industry. In other common cases crucial public goods (such as knowledge, information and know-how) may be underprovided owing to non-rivalness, non-excludability and the free-rider problem. In these cases policy intervention may help the productive system to follow more desirable and efficient paths.

The second typical rationale in favour of government intervention refers to criteria that go beyond efficiency (see Figure 0.1). Experiences around the world suggest that there are additional justifications based on strategic objectives (such as employment and national competitiveness) and based on what we may define as *meta-economic objectives* (such as distribution of wealth and economic opportunity) (Chang, 1994; Amsden, 1989).

In this context, there are lessons to be learned from the experiences and practices of successful economies like France, Italy, Japan, Korea and China. In many cases in these countries, industry has been promoted for reasons that go beyond efficiency criteria and the correction of market failures. In many of these cases industry has been promoted for strategic reasons because government has wanted to protect infant industries, maintain declining sectors, or support 'national champions'. In the past, strategic sectors have included motor cars and steel. Today, the sectors that are considered strategic in all highly industrialized countries (and some emerging countries) include new industries such as electronics, telecommunications and biotechnology. In these cases, justifications for governmental intervention go beyond issues of market failure discussed above. However, again, if we analyse the *histories* of industrial development it is clear that in many countries – developed, developing and in transition – industries have also been favoured or promoted for reasons that we may define as either paternalistic or 'meta-economic', meaning beyond traditional

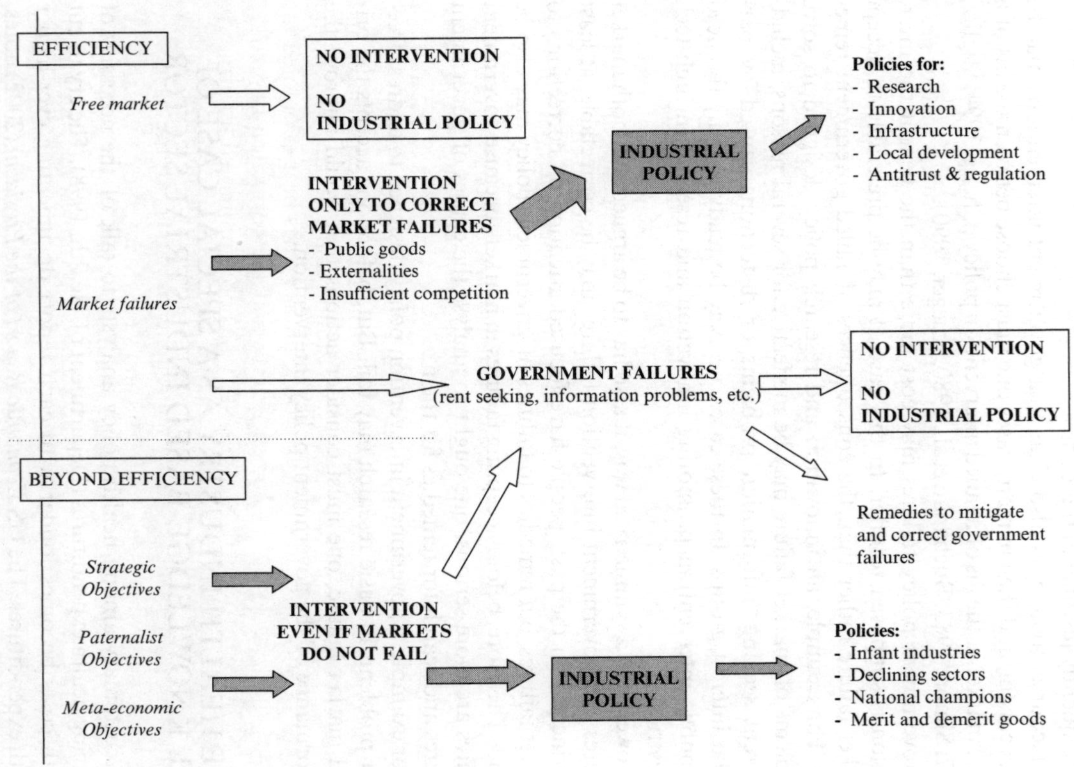

Figure 0.1 The industrial policy debate

economics rationales. Here industry is seen as an instrument to achieve wider goals, including access to knowledge and health, improved distribution of wealth, and social or environmental sustainability. Government promotion of hi-tech sectors such as telecommunications and biogenetics are frequently justified in these ways.

Of course, today it is also clear that government interventions may be unsuccessful, and then in many cases one must choose between the ills of a failed market or the ills of a failed intervention policy (Chang, 1994; Stiglitz, 1987; Simon, 1983; Buchanan et al., 1980; Kueger, 1990).

Government policy failures may cost more than the benefit the intervention is supposed to offer. In this case, it may be preferable to accept market failure rather than the consequences of failed government remedies. For example, an innovation and research policy designed to solve problems of market failure may be ineffective for several reasons, including rent seeking, information problems or distortions caused by particular interest groups. In these cases, we may be ready to deny the need for public intervention promoting innovation and research in high-tech industries.

However, in a symmetrical way, it also has to be argued that both market failures and government intervention failures may be correctable, at least in principle. In the past, people have focused attention on corrections to market failures, but remedies to failures of government policy may also be found. Therefore, before accepting failures in markets because government failures are more serious, one ought to analyse the causes of government failures and search for remedies for them.

For example, a government intervention policy designed to address free-rider problems in basic research may fail. But, before one accepts the original market failure, one must consider actions that would improve the performance of the government policy intervention.

THE HEALTH INDUSTRY AS A SPECIAL CASE OF THE KNOWLEDGE-BASED INDUSTRIAL SECTOR

It is customary among health policy analysts to talk of the necessity of reducing health expenditures (Department of Health, 1998). The cry to cut health costs becomes louder with each report documenting ever-higher health expenditures. The US *Economic Report of the President* (2002) notes, for example, 'There is . . . growing evidence that substantial opportunities remain both to reduce costs and to achieve greater improvements through more effective use of medical services – that is, to improve the value, or output per dollar spent, of our health care system'. The rationale for this urge to reduce health expenditures is the assumption that the objective of a nation's health system is to produce an acceptable level of health at minimum cost. We call this objective 'the Health Model'. The Health Model portrays health services as an input in a production function that has health as the only output. The responsibility of the health system is to produce an acceptable level of health as efficiently as possible (see, for example, Cochrane, 1972). The task is difficult because it is not easy to determine what the 'acceptable' level of health is for a particular population. People's preferences differ widely in terms of health and other determinants of consumer satisfaction, and some contend that it is in fact impossible to aggregate personal utility functions to determine what society prefers. By stating that the only objective of the health system is to produce health, it is reasonable to demand efficiency in the system, such that the acceptable level of health be produced with the minimum level of expenditures.

Another view of health systems, however, considers other objectives than health status. This other view suggests that the health system also produces benefits for an economy in terms of economic externalities that result from investment in research and development, and it can produce services that will be sought by the world economy and hence produce valuable export earnings. We call this view 'the Health Industry Model'. The Health Industry Model views the health industry's outputs in the framework of a multi-product production function, in which health and other outputs are produced by health providers and health manufacturers. The Health Industry Model suggests that the health sector comprises a large number of interconnected industries that produce both health services and manufactured goods, including pharmaceuticals and diagnostic equipment. The health industry is not the only industry that produces broader societal benefits in addition to its stated mission. Educational systems, for example, produce broader benefits than knowledge, spilling over to improve labour productivity, political awareness and cultural enrichment. But the health industry is particularly interesting in being able to produce broader technological development, as well as health, because of its high-tech nature. The Health Industry Model is described in detail in the first chapter of this volume, by Di Tommaso and Schweitzer. The health industry contains three clusters of industries: health care providers, health care financiers, and health manufacturers. The structure of the health industry is shown in Figure 0.2.

Health care providers consist of both institutional and community-based organizations. Institutional organizations are hospitals, nursing homes, mental hospitals and so forth, while community-based organizations include ambulatory care clinics, community-based medical practices,

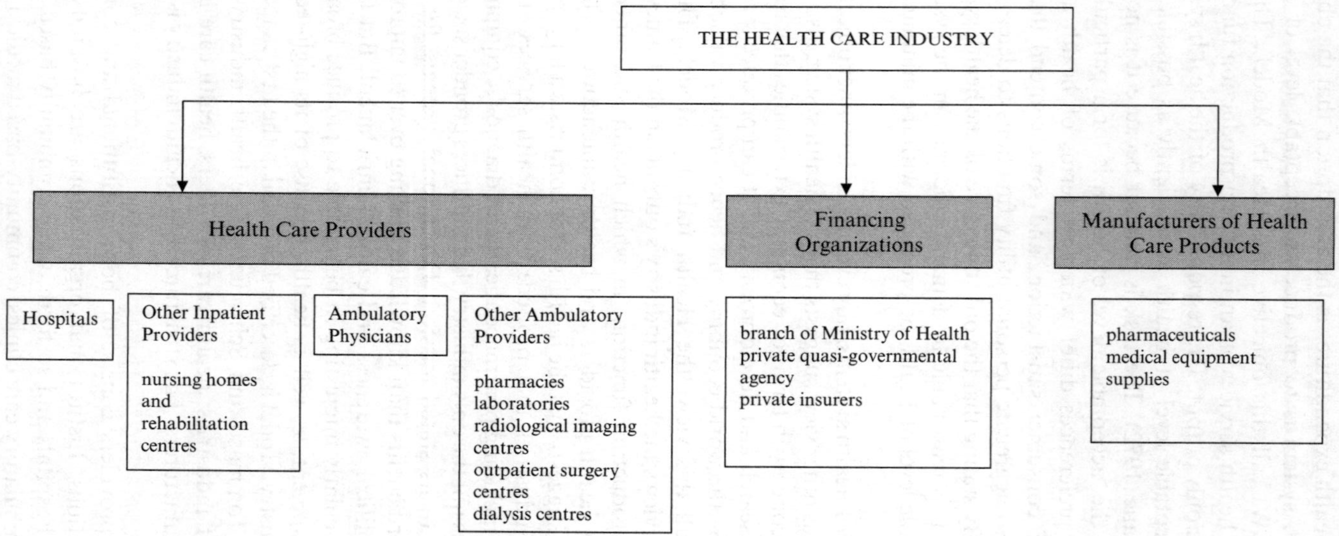

Figure 0.2 The 'health industry' model

home-based care programmes such as Visiting Nursing and home-based hospice services. The ownership of institutions varies from country to country. In some countries, such those in Europe, most hospitals are government-owned, while, in the USA, most hospitals are private, though some are owned by state or local governments. In most countries community-based physicians are independent entrepreneurs, though in some countries they are civil servants, employed by government. In recent years there has been a substantial shift in the locus of care from institutional providers, such as hospitals, to home-based care programmes.

Health care financiers are similarly either public or private. Most countries have health care financial systems that are either public or quasi-public. The USA is remarkable for funding over half of national health expenditures through private health insurance firms. Public expenditures in the USA, however, are very large, consisting primarily of the Medicare programme for the elderly and the 50 state Medicaid programmes for the indigent (*Health United States: 2003*). In other countries, though public financing of health care predominates, private insurance coverage is expanding and the share of privately financed health care is increasing. Therefore, though the relative proportions of public and private financing differ between the USA, European countries and Japan, most financing programmes in industrialized countries are, in fact, mixed, with public and private funding coexisting.

The third group of industries in health care comprises health product manufacturers. These firms are well known for manufacturing pharmaceuticals and diagnostic and therapeutic medical equipment, such as radiology and dialysis machines, and laboratory testing equipment.

Two important observations about the health industry follow from this discussion. The first is that the outputs of the health industry are varied and complex. In contrast to most industries, the outputs are both products and services. This is a major reason why cost containment is so difficult in health, because the production process in that part of the industry that is producing products is relatively transparent, while the production process for services is less so. Inputs are difficult to quantify and so are outputs, because quality, as well as quantity, are important dimensions of output.

The second observation is that virtually all of the health industry's components are interconnected, forming a complex system of providers, insurers and manufacturers. This is why efforts to reduce overall health expenditures by reducing expenditures in one industry (for example, hospitals) so often fail. Decreases in one industry's expenditures are likely to lead to increases in expenditures of other industries (for example, pharmaceuticals and ambulatory care). And when administered prices in one industry are reduced (for example, by lowering physician reimbursement rates), increased utilization of substituting services, including hospital care, emergency room visits and pharmaceuticals, are likely to result. It is often observed that the pharmaceutical industry, for example, both substitutes for other health services, and complements them (Comanor and Schweitzer, 1994).

The objectives of the health industry are difficult to ascertain. While most health systems are regulated according to the traditional view that a health system's objectives are to produce an acceptable level of health for a population at minimum cost, the health system is more complex than this simple mission would suggest. In addition to maintaining a population's health status at some acceptable level, the health system is a major generator of scientific knowledge through its research and development functions. Increases in health status produce not only social welfare, but also increased productive capacity of a nation's workforce. This effect of health status on economic development in developing countries has been noted by the World Health Organization (World Health Organization, 2002). The effect of health on both the educational system and the labour force applies as well to industrialized countries, but the benefits of a strong health industry are not limited to direct health effects. Scientific knowledge and productive capacity produce benefits throughout the economy. Therefore these benefits are described as 'public goods' and, if left to individual markets to produce, they will be underproduced. No firm has a sufficient incentive to invest in knowledge or worker health, because the benefits extend far beyond the single firm. The health industry is also capable of producing other benefits, such as 'national champions', industries in which a nation can take pride, especially when this industry represents new technological breakthroughs. A last consideration is that high-technology industries typically start out small and need 'nurturing'. With sufficient protection and public investment, there is the hope that industries will grow to be self-sustaining. Examples include biotechnology and the new field of nanotechnology (the production of ultra-small structures that can be used to construct very strong materials or devices). The existence of a wide range of health industry outcomes suggests that government has an important role in funding those functions of the health industry that produce these broad benefits.

Understanding the multi-objective role of the health industry explains, in part, why policy decisions affecting the size of the health industry are often so contentious. There is a conflict between those who see the health industry as responsible only for producing an acceptable level of health (at minimum cost) and those who see broader benefits of a dynamic, scientifically sophisticated health industry. The perception of broader benefits from the health industry justifies increased government investment to address market failure concerns and other objectives of a country's health industry.

Government policies towards the health industry can be described in terms of the degree of central authority inherent in a policy. National policy can be characterized as a 'macro' activity: the set of policies designed to affect an entire economy. Policies can also be directed at the level of the firm itself. The focus of these policies can be called 'micro'. Then there is a level of policy analysis that is neither macro nor micro, but rather based upon agglomerations of firms and institutions, usually within a single industry, though not necessarily so. These agglomerations are referred to as 'industrial clusters'. Considerable interest is expressed these days in developing policies that will encourage the establishment and growth of these industrial clusters, especially those that concentrate in high-technology industries such as biotechnology.

The book follows this three-part framework in focusing on industrial policy in the health industry. The first chapter, by Di Tommaso and Schweitzer, describes a new view of a nation's health system and justifies an increased role for government in promoting the health industry. The traditional view of a health system, called 'the Health Model', is that its objective is to produce an acceptable level of health for its population at minimal cost. The new view, called 'the Health Industry Model', suggests additional policy objectives, most prominent of which is the production of knowledge and innovation that will spillover into other industries. The implication of this model is that many health system activities are in the realm of public goods and must be financed by government rather than by individual providers of manufacturing firms.

POLICIES TO ENCOURAGE DEVELOPMENT AND ADOPTION OF NEW TECHNOLOGY

Three Perspectives: Macro, Micro and Intermediate-Level

In the remaining chapters of this volume we look at industrial policy in the health industry from three viewpoints. The first, the economy-wide macro view, looks at the effect of government policies in encouraging innovation and high-tech industrial development. The second view, the micro view, discusses both production of innovation and adoption of technological change by individual health industry actors: firms, universities, hospitals, and so on. Finally, we consider the intermediate level, which is neither economy-wide nor specific to a single actor: the industrial cluster, defined as a geographic agglomeration of firms and institutions operating within a single industry or a group of complementary sectors.

The 'Macro View'

How can governments encourage innovation, diffusion of knowledge and adoption of new technologies with broad policies? Three chapters in this volume deal with these macro-oriented policies.

In Chapter 2, Branston, Rubini, Sugden and Wilson are concerned that broad strategic health policy decisions may be made, not in the interest of a nation's constituency, but rather by elite groups within each firm in the industry. One implication of this may be that decisions concerning investment in innovation in the health industry may reflect the wishes of administrators and managers rather than the needs of the population. Two levels of failure occur in this scenario. The first is that single health providers may not adopt innovation or improve quality to the desired degree because provider objectives replace public objectives. The second is that the industry, without coordination, will produce less innovative and lower-quality health care than would be desirable to the broader constituency.

Chapter 3, by Vicarelli, poses three alternative strategies by which governments have attempted to contain health care expenditures. The first is a top-down approach of curtailing demand by restricting supply. The second approach introduces competition between providers in an attempt to increase provider efficiency. The third approach, initially implemented in Scandinavia, introduces cooperation and collaboration, rather than competition. In this approach providers take a more active role in defining regulation. In describing these three cost-containment strategies, the author provides a framework to assess which approach is best able to promote innovation and improve quality of care.

In Chapter 4, Comanor, Schweitzer and Carter look at another governmental policy area (price setting) as a determinant of research and development and innovation. The setting of this empirical study is the US pharmaceutical industry. In looking at pharmaceutical reimbursement, the study explores the relationship between market factors, product characteristics and prices for pharmaceuticals. The finding is that product characteristics – efficacy, safety and convenience – are important determinants of market prices of drugs. The implication of this finding for explaining variation across countries in investment in R&D is great, because the evidence is that consumers have strong preferences for drugs that excel along each of the three dimensions. Attempts to restrict prices for products that improve efficacy, safety or convenience will restrain investment in innovation and consumer welfare will be reduced.

The 'Micro View'

Macro policies to facilitate innovation and technological change are important, but are insufficient in themselves. Individual institutions must be willing to innovate and accept change. We present three chapters that look at the way particular actors in the health industry were able to adapt their structures to react to innovation and change. With few exceptions, development of new technology depends both on the system's capacity to produce innovation and on its ability to adopt, or absorb, new technology. A system characterized by actors that are unable to accept technological change will discourage innovation in the first place.

Vagnoni, Guthrie and Steane (Chapter 5) study the pivotal role of universities in developing new technology. The authors are particularly interested in university relationships with industry in sponsoring research. While acknowledging the vital role of the industrial sector in collaborating with universities, they point out that underlying core funding by government for basic research is essential for production of innovation.

The chapter by Bianchi and Labory deals with the concept of intangible assets in the pharmaceutical industry. The authors consider the role of spillovers in high technology industries, generally, consistent with the Health Industry Model. They suggest that it is the intangible assets, whose share of industrial asset value has increased dramatically in recent decades, that are largely responsible for the spillovers that occur between high-tech industries. They discuss the nature of intangible assets and how this particular type of asset affects a firm's ability to innovate and increase productivity and value.

In Chapter 7, Llewellyn and Northcott focus on government policy directed towards improving efficiency in hospitals. The authors analyse a UK initiative that publishes hospital cost data so that individual hospitals can see where they excel and where they are deficient in comparison to others. They argue that facilitating these comparisons could, in theory, improve provider efficiency by encouraging the implementation or adoption of efficiency-increasing innovation. On the other hand, interpreting and making use of these comparative cost data is more complex than might be expected and the UK experience shows that this attempt to increase hospital efficiency was less successful than was hoped.

The 'Intermediate View'

In-between macro policies and firms themselves, there is another emerging focus of industrial policy that is neither at the level of the overall economy nor at the level of individual firms. This is the agglomeration of firms and institutions within a geographic area, usually within a single industry, into high-technology industrial clusters. In many cases throughout the world, these clusters are the direct result of government policies. Other clusters, however, are the natural result of market forces, without government intervention.

It is often noted that industrial clusters have existed for centuries (Becattini, 1990). Furthermore recent literature in the field of industrial organization has noted the importance of these clusters in increasing industry efficiency and competitiveness. Most of this literature, however, refers to the industrial sector and not to high-technology industries. Some have suggested that the nature of high-technology industries (for example, the lack of need for physical proximity to inputs or markets) makes clustering in this sector obsolete (distance is obsolete). The evidence, however, is strongly to the contrary, and we and our colleagues point out that these clusters are as central to the development of 'new', high-technology industries as they were to manufacturing industries.

In the final part of this volume we present five chapters describing high-technology industrial clusters in several countries. These chapters explain how it is that clusters arise, and what industrial policies appear to facilitate their appearance in different countries.

Chapter 8, by Di Tommaso, Paci and Schweitzer deals with clusters of firms characterized by intangible assets. An important example of this is the biotechnology industry. While one might expect that modern information technology would make spatial agglomeration of these firms unnecessary, the authors describe the particular rationale for these firms to cluster.

In Chapter 9, Schweitzer, Connell and Schoenberg develop further the ideas of the previous chapter. They present empirical evidence on factors that are associated with the existence of biotechnology clusters in the USA. The implications of this analysis are that universities are instrumental in being 'seed-beds' of biotech firm development, and that policies that stimulate basic research in universities are likely to lead to development of high-tech industrial clusters.

Chapter 10, by Hirsch, attempts in an innovative way to calculate the economic spillover effects of a university-sponsored industrial cluster on its region. A three-stage model is proposed and tested with reference to a large biotechnology cluster in California. The first stage of benefits is the direct university expenditure on goods and services in the cluster. The second, indirect, stage is the effect of these expenditures on the purchase of goods and services by those firms. The third stage consists of expenditures by employees of these firms. These three stages of benefits are a lower estimate of total benefits because additional indirect effects exist, but are even more difficult to measure. One of these is the effect of externalities and synergies by which growth of one industry produces benefits for other industries, either in the same region or further away.

Bellandi and Tessieri present a chapter dealing with relationships between Italian high-tech firms operating within a cluster and foreign multinational firms. The authors demonstrate that multinational firms will often seek existing industrial clusters in which to locate their firms or invest capital in order to exploit advantages and economies that they would otherwise be unable to achieve. These foreign-based firms then develop strategic linkages and alliances with the Italian firms already in the cluster.

Finally, Chapter 12, by Katz-Bénichou and Viens describes the development of two high-tech clusters in France. The authors point to an entirely different model of high-tech cluster development, suggesting that countries may have different ways of promoting industrial development.

NOTES

* We wish to thank Patrizio Bianchi (University of Ferrara and L'institute), Roger Sugden (University of Birmingham and L'institute) and William Comanor (UCLA) for their stimulating comments to a preliminary draft of this introduction.

1. Interventions can be direct or indirect. The *Economic Report of the President* (2002), for example, refers more to 'incentives' than to direct intervention in stating, 'It is important to establish incentives that will ensure continued growth in innovation and the new technologies that will define the 21st century. We must not only invest in basic research, but also ensure that intellectual property of innovators is secure at home and abroad' (p. 62).
2. As Gerosky pointed out: 'any random collection of six economists is sure to produce at least a dozen different opinions on the subject, not least because many economists have trouble in reconciling their gut reaction that industrial policy should not exist with the obvious fact that it does' (Gerosky, 1989, p. 20).

REFERENCES

Amsden, A. (1989), *Asia's Next Giant: South Korea and Late Industrialization*, Oxford: Oxford University Press.

Amsden, A. (1994), 'Why isn't the whole world experimenting with the East Asian model to develop? Review of the East Asian miracle', *World Development*, **22**(4).

Becattini, G. (1990), 'The Marshallian industrial district as a socio-economic notion', in F. Pyke, G. Becattini and W. Sengenberger (eds), *Industrial Districts and Interfirm Cooperation in Italy*, Geneva: International Institute for Labour Studies.

Buchanan, J., Tollison R. and Tullock, G. (eds) (1980), *Towards a Theory of Rent Seeking Society*, College Station, Texas: Texas A&M University Press.

Chang, H.J. (1994), *The Political Economy of Industrial Policy*, London: Macmillan.

Cochrane, A.L. (1972), *Effectiveness and Efficiency: Random Reflections on Health Services*, Oxford: The Nuffield Provincial Hospitals Trust.

Comanor, W.S. and Schweitzer, S.O. (1994), 'Pharmaceuticals', in Walter Adams and James Brock (eds), *The Structure of American Industry*, 9th edn, New York: Prentice-Hall.
Cowling, K. (ed.) (1999), *Industrial Policy in Europe*, London and New York: Routledge.
Deane, P. (1989), *The State and Economic System*, Oxford: Oxford University Press.
Department of Health (1998), *The New NHS – Modern and Dependable: A National Framework for Assessing Performance*, Leeds: National Health Service Executive.
Economic Report of the President (2002), Washington, DC: United States Government Printing Office.
Gerosky, P. (1989), 'European industrial policy and industrial policy in Europe', in *Oxford Review of Economic Policy*, **5**(2).
Kueger, A. (1990), 'Government failure in economic development', *Journal of Economic Perspectives*, **4**(3).
Musgrave, R. and Musgrave, P. (1984), *Public Finance in Theory and Practice*, New York: McGraw-Hill.
Simon, H. (1983), *Reason in Human Affairs*, Oxford: Basil Blackwell.
Stiglitz, J. (1987), 'The principal agent problem', *The Palgrave Dictionary of Economics*, vol. 3, London.
Stiglitz, J. (1988), *Economics of the Public Sector*, New York: W. Norton.
US Department of Health and Human Services (2003), *Health United States, 2003*, Washington, DC: US Government Printing Office.
World Health Organization (2002), *Macroeconomics and Health: Investing in Health for Economic Development*, Geneva: World Health Organization.

1. The Health Industry Model: new roles for the health industry

Stuart O. Schweitzer and
Marco R. Di Tommaso

INTRODUCTION

There is wide variation in technological development and innovative capabilities of industrialized countries. What explains these differences in the pace of innovation and dissemination of new scientific knowledge? What are the *policy* actions that might be undertaken if governments wanted to stimulate technological innovation? To what extent are these actions consistent with traditional health policy approaches? Can the health industry be used as a leading sector, stimulating other high-tech industries?

In this chapter we suggest that answers to these complex questions can be suggested by seeing the health care sector from a new perspective. The health industry is one of the largest industries in any wealthy and industrialized economy, measured in terms of expenditures and employment. The industry's size is not its only characteristic, however. Technologically the health industry is central to other high-tech or 'new' industries. Therefore government policies affecting the health industry will have widespread effects in other technologically sophisticated areas. This new perspective suggests a rethinking of the definition of the public policy tools and objectives to be applied in a sector which, over the past few decades, has been strongly influenced by two factors: budget constraints and technological progress.

In these countries most health care demand is financed by public resources. This characteristic has pushed policy makers to focus their attention on the opportunity costs of health spending and in this context health expenditures have been strongly limited by more general public policy objectives. As a result, many governments have succeeded in restraining the level and proportion of health expenditures with mechanisms designed to improve the efficiency of publicly financed health services and to orient demand towards private suppliers.

Despite these efforts, spending on health has continued to grow in all industrialized countries and will no doubt continue to rise, largely owing to three interrelated factors: the development of new technologies that offer a wider range of treatment options to patients, demographic trends and greater expectations on the part of consumers, who now more than ever actively seek information about the goods and services that are available.

We begin our analysis of the health sector in a traditional way by describing its magnitude and structure. We will then introduce our alternative perspective in order to understand better the components, weight and potential of what might be defined as the 'health industry'. Finally, we turn our attention to the policy implications suggested by this new perspective.

THE SIZE OF THE HEALTH SECTOR

Health Expenditures

A first indicator of the size of a country's health sector is the proportion of its GDP allocated to health goods and services. Table 1.1 shows the magnitude of the health sector in the G-7 countries from 1990 to 2000 according to GDP share.

This measure varies widely across the seven countries included in our comparison, from a low of 7.3 per cent in 2000 in the UK, to a high of 12.1 per cent in the USA. More typical among European countries are the figures of France and Germany, which are around 10 per cent. Though most OECD countries have experienced an increase in the share of GDP allocated for health, a few have experienced no growth from 1995 to 2000 (Germany and Canada) or even a slight reduction in share, such as France. Data for three years are presented because the GDP share is a ratio and is

Table 1.1 Share of GDP (%) allocated to health in the G-7 countries

	1990	1995	2000
Canada	9.0	9.2	9.2
France	8.6	9.5	9.3
Germany	8.5	10.6	10.6
Italy	8.0	7.4	8.2
Japan	5.9	6.8	7.6
United Kingdom	6.0	7.0	7.3
United States	11.9	13.3	12.1

Source: OECD (2003).

Table 1.2 Per capita health expenditures in the G-7 countries (US dollars)

	1990	1995	2000
Canada	1676	2114	2535
France	1517	1980	2349
Germany	1600	2264	2748
Italy	1321	1486	2032
Japan	1083	1631	2012
United Kingdom	972	1315	1763
United States	2739	3703	4631

Source: OECD (2003).

likely to vary according to the stage of the business cycle. Because the level of expenditures (the ratio's numerator) is rather stable throughout the cycle, the ratio will vary inversely with the level of GDP (the ratio's denominator). Thus, in an economic expansion, the GDP share will tend to fall, while in a recession it will rise.

Per capita health expenditure is another indicator of the size of the health sector in comparative analyses, and avoids the simultaneous influence of size of GDP. Table 1.2 shows per capita national health expenditures for the G-7 countries expressed in US dollars, adjusted for purchasing power parity (PPP).

All countries have experienced a remarkable increase in health expenditures from 1990 to 2000 and such expenditures are substantial even for the most frugal of the countries, the UK, Japan and Italy, whose annual per capita expenditures are $1763, $2012 and $2032, respectively. Germany and France, spending $2748 and $2349, respectively, are more typical of European countries, while the USA, by far the most profligate health spender in the world, spends $4631 per person.

Some argue that the high expenditures in the USA result from inefficiencies inherent in a competitive marketplace, that lead to higher administration and marketing costs, higher labour costs, redundant capital equipment and unnecessary utilization (Rice, 1997, 1998; Evans and Roos, 1999). An alternative explanation is that US health expenditures are driven by consumer choice working through a more private-oriented market-driven financing system (Enthoven and Singer, 1997).

Health Sector Employment

Another measure of the size and importance of the health sector is its employment, as shown by Table 1.3. Even for those countries whose

Table 1.3 Health services employment and share of total employment in the G-7 countries, selected years

	Employment person man-years	Share of total employment (%)
Canada (1994)	731 700	5.5
France (1994)	1 522 319	6.9
Germany (1995)	2 325 000	6.7
Italy (1992)	1 023 500	4.8
Japan (1993)	2 380 093	3.7
United Kingdom (1995)	1 182 000	4.6
United States (1993)	8 414 000	7.0

Source: OECD (1998).

Table 1.4 Employment in selected sectors in the G-7 countries, 1996[a]

	Health services	Food products	Textiles	Industrial chemicals	Transportation equipment	Total manufacturing
Canada	732[b]	191	63	28	199	1 691
France	1 522[b]	141	147	98	482	3 789
Germany	2 325[c]	442[b]	189[b]	287[b]	865[b]	6 731[b]
Italy	1 024[d]	184[b]	210[b]	71[b]	283[b]	2 822[b]
Japan	2 380[e]	1 171	456	152	888	10 287
UK	1 182[c]	455	182	102	409	4 163
USA	8 414[e]	1 396	820	363	1 689	17 248

Notes: [a] all data are in thousands, and pertain to 1996 except where otherwise noted; [b] 1994, [c] 1995, [d] 1992, [e] 1993.

Source: OECD (1998) and UNIDO.

expenditure shares on health are the least (UK and Italy) the employment share is substantial, at nearly 5 per cent. For countries spending more (France, Germany and the USA), the employment share approaches 7 per cent.

Table 1.4 shows health services employment in relation to employment in other selected sectors and industries for the G-7 countries. Health sector employment exceeds the other industries in the table, and is typically from 25 per cent to 50 per cent as large as employment in the entire manufacturing sector.

As impressive as the health services employment figures are, the figures in Tables 1.3 and 1.4 undercount employment because they are restricted to 'Health services', and therefore exclude workers employed in the pharmaceutical, biotech, and medical equipment manufacturing industries. As we will develop further in the following pages, we define the 'health industry' to include all the related industries, each of which contributes (directly or indirectly) to satisfy the 'demand for health'. Hence, from this perspective, health spending as a percentage of GDP (Table 1.1) and health expenditures (Table 1.2) better represent the magnitude of what we call the health industry.

TWO MODELS OF THE HEALTH INDUSTRY

For analysis and policy development purposes, we propose two models that describe health policy and the functioning of the health sector: the 'Health Model' (HM) and the 'Health Industry Model' (HIM). The former, concentrating on health outcomes, has been the usual way of looking at health care for decades. But recent trends towards high rates of research and development in the industry, more sophisticated and advanced technologies, use of highly trained personnel and the opening of international boundaries to the flow of health care services suggest the need for a new analysis and policy approach. We propose the Health Industry Model, the grouping of related health care activities, as one of the major 'industries' in advanced economies.

The Health Model (HM)

In the traditional Health Model, the focus of policy analysis and debate is on how to produce an acceptable level of health for the population within a constrained budget.

An acceptable level of health

The first issue in the HM is defining what is meant by an acceptable level of health. This question is multidisciplinary in nature and beyond the scope of this chapter. However, a few considerations are important.

The problem is difficult because consumers differ in their preferences for various health states and so the aggregation of personal utility functions is at best difficult and, according to some, impossible (see Arrow et al., 1996). An example of the dilemma faced by health systems is choosing the allocation of resources among various health services. For example, if the health system attempts to minimize mortality rates subject to available resources, treatment of non-life-threatening cases will tend to be underfunded.

It is important to note that neither health spending nor resource supply is an adequate indicator of the 'acceptability' of the level of health. For example, considering health expenditures or the number of physicians per 100 000 people as an indicator of the acceptability of health would be to confuse inputs with outputs. Sen (1999a, 1999b) refers to a measure of health outcomes as the 'capabilities' of populations to lead long and healthy lives. These capabilities are more than health status alone, and are dependent on the structure of *entitlements* in the society, the set of rules, both formal and informal, that define access to health, such as education, social norms and so forth.[1] Whether health systems, in theory, can ever meet the expectations and aspirations of the populations they serve is, perhaps, a less serious issue in practice in democracies because governments must satisfy the demands of a majority of their populations in order to be elected. Of course competing issues (such as defence, taxes and foreign affairs) attenuate the importance of any particular issue (such as health). Nonetheless health is a high enough priority for a government that fails to address the issue probably not to survive in the long run.

Another way of looking at the complex relationship between health inputs and health capabilities is the *Health Behavior Model* of Anderson (1995), that categorizes determinants of health services utilization into groupings of predisposing, enabling and need variables. This model has been useful in explaining why utilization of health care differs among groups according to social class, education, income and other variables (UNDP, 1999).

Wide variation exists in the amount of resources devoted to health care and of the amount of health goods and services produced, even across members of the G-7 countries. Similarly wide variation exists in health status and capabilities, though the association between inputs and outputs is far from clear. The data in Table 1.5 portray examples of resource availability (such as per capita health expenditures and physicians per 1000 people) and two indicators of health status (or output or capabilities): infant mortality and life expectancy for males. As shown by the table, it is clear that it is very difficult to quantify *capabilities* precisely and these difficulties do not justify the confusion between input and output, as above mentioned (UNDP, 1999).

Functioning within constrained budgets

As already mentioned, the health care cost-containment objective has been, and still is, a major concern in all industrialized countries. This is particularly true in the European countries, in which health systems are primarily financed with central government direction, sometimes called 'top-down' budgeting. In the G-7 countries, with the exception of the

Table 1.5 Selected health service inputs and capabilities among the G-7 countries, 2000

	Inputs		Capabilities	
	Per capita health expenditures	MD per 1000 people	Infant mortality deaths/1000 live births	Life expectancy, males
Canada	2535	2.1	5.3[a]	76.7
France	2349	3.3	4.5	75.2
Germany	2748	3.6	4.4	74.7[a]
Italy	2032	6.0	5.1	76.3
Japan	2012	1.8	3.2	77.7
UK	1763	1.8	5.6	75.4
USA	4631	2.8[a]	7.1[a]	74.1

Note: [a] 1999.

Source: OECD (2003).

Table 1.6 Public share of health expenditures among the G-7 countries

	1990	1995	2000
Canada	74.5	71.4	70.9
France	76.6	76.3	75.8
Germany	76.2	76.7	75.0
Italy	79.3	72.2	73.4
Japan	77.6	79.1	78.3
United Kingdom	83.6	83.9	80.9
United States	39.6	45.4	44.2

Source: OECD (2003).

USA, the government share of health care funding in 2000 ranged from 71 to 81 per cent, with a declining trend in nearly all countries, as shown in Table 1.6.

The public share of health expenditures has been remarkably stable over the past few decades, as shown in Figure 1.1. Exceptions were the major shifts in health financing that occurred in both Canada and the USA in the 1960s, substantially raising the public share of health expenditures in both countries. The change in Canada was more comprehensive, with its share rising to those of the European countries and Japan.

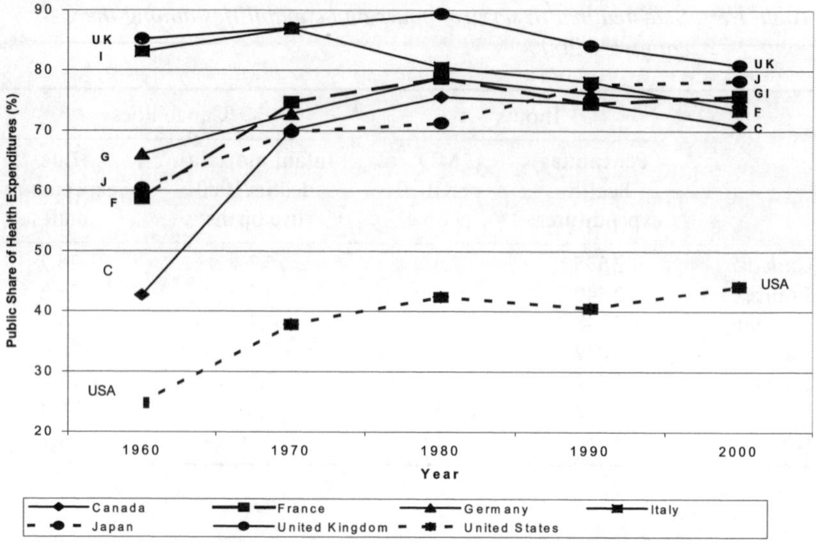

Source: OECD (2003).

Figure 1.1 Public share of health expenditures in the G-7 countries, 1960–2000

It is recognized that the high levels of health spending limit expenditures on other important priorities, including education, urban infrastructure and rural development. Health spending is generally determined by public policy, assuming an insatiable demand for health care, and inefficiencies in supply.

The American health care system works differently from European systems because of its high level of private expenditures (nearly 56 per cent of total health expenditures in 2000). Because of the high proportion of private health expenditures, consumption is more likely to reflect individual consumer preferences than would expenditures determined directly by government policies, as in the case of Europe.

The USA is unique in having multiple health plans within most geographic areas, each with different insurance coverage and provider access. Under these competitive conditions, consumers can choose health plans according to price and quality. Inefficient health plans – those that provide coverage that consumers do not want, those whose costs are excessive because of high provider payment rates or failure to screen out unnecessary utilization, or those that allocate too great a share of premium income to overheads and too little to reimbursement for care – cannot survive in a competitive atmosphere in the long run. The ability of competition to achieve market efficiency depends in part on the ability of consumers to obtain sufficient information about health care plans and providers. This is the context of current efforts by large employer groupings in the USA, and the US Medicare and Medicaid Services Administration to increase consumer information by producing 'report cards' that rate providers and insurance plans. The attempts, especially by governments, to produce an acceptable level of health within a constrained budget (and constantly to look for ways of lowering health care costs) is the essence of the Health Model.[2]

The Health Industry Model (HIM)

In the Health Industry Model, attention is focused on broader potentials that the health industry can supply to advanced economies. This approach views the health sector as a broad industrial grouping including constituent parts such as the pharmaceutical industry, the biotechnology industry, the medical equipment industry, telemedicine and so on. The relationship between the components of the health industry is shown in Figure 1.2, which shows that the health industry consists of three main components: the providers of health care, financing organizations and manufacturers of health care products.

Health Care Providers

There are large differences among countries in the relative importance of providers, as measured by expenditure share. Table 1.7 shows these shares for two providers: hospitals and outpatient facilities.

Hospitals are the largest category of health care provider, accounting for more than one-third of all health expenditures in four countries included in our analysis and nearly 30 per cent in two others. In Italy and France, this share exceeds 40 per cent of total spending. In Canada and the USA, the 2001 hospital share is unusually low and results from a particularly large decrease from the 1995 share. There is more variation in ambulatory physician service expenditures between countries. Even excluding Japan, ambulatory physician services account for from under 12 per cent of total health care expenditures in France, to close to 20 per cent in the USA.

There is a theoretical reason for suggesting either a substitution effect between hospital and outpatient services (in which a higher share in one sector leads to a lower share in the other) or a complementary effect (in which the shares move in the same direction). After looking at the data, however, neither effect appears to be consistently evident.

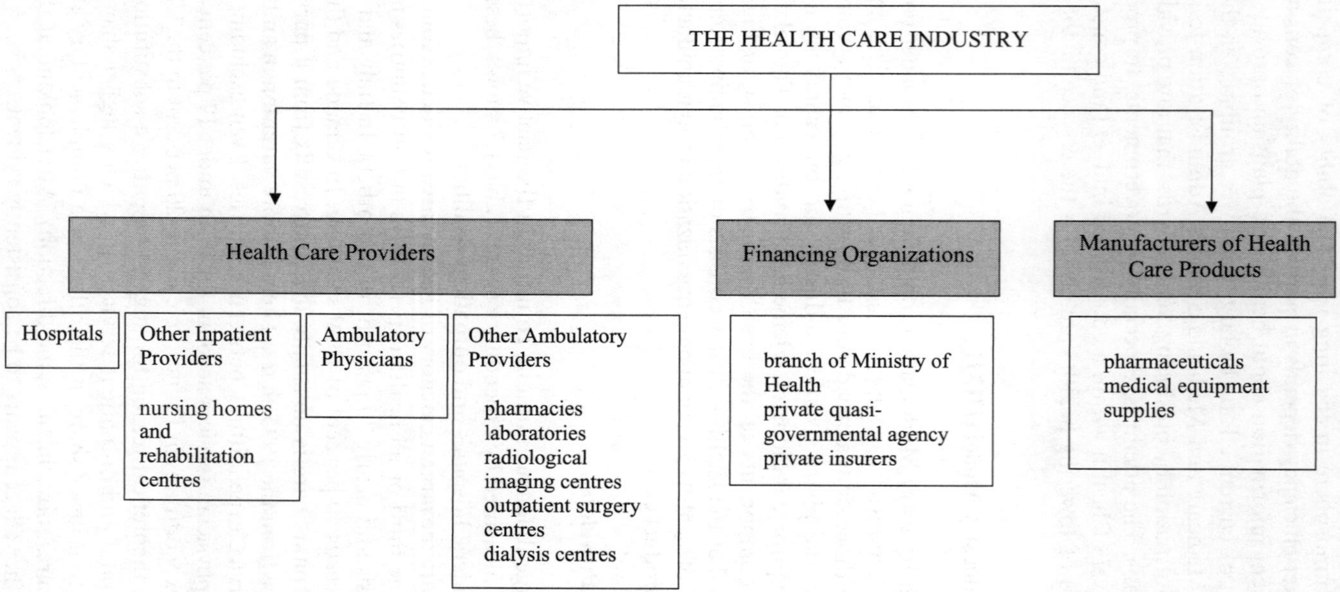

Figure 1.2 The 'health industry' model

Table 1.7 Proportion of national health expenditures spent on selected health care providers: inpatient hospital services and ambulatory physician services in the G-7 countries

	Inpatient hospital services (%)		Ambulatory physician services (%)
	1995	2001	1995
Canada	44.7	29.8	14.5
France	45.1	41.6	11.7
Germany	36.9	36.1	16.6
Italy	44.8	41.1	20.8
Japan	36.4	37.9[a]	34.3
UK	n.a.	n.a.	15.2[b]
USA	32.2	27.1	19.8

Notes: [a] 2000, [b] 1994.

Source: OECD (1998, 2003).

Hospitals

Hospitals are organized differently in different countries. In many countries of Europe, including Italy, the UK, France and Germany, most hospitals are owned by the government or by the national health service. Even in these countries, however, some hospitals are privately owned and serve patients seeking care outside the state health system (and some beds in public hospitals may be set aside for private patients). In the USA, ownership is different, with most hospitals owned by not-for-profit local organizations ('community' hospitals). Additionally an increasing number of hospitals in the USA are owned by investors, and are for-profit entities, as shown in Figure 1.3.

While the *share* of not-for-profit beds grew between 1980 and 2000, their number actually fell by 16 per cent from 699 000 to 588 000. The number of investor-owned beds grew by nearly 32 per cent, from 82 000 (6 per cent of the total) to 108 000 (11 per cent of the total).

This growth of investor-owned hospitals in the USA has been particularly rapid in the past 20 years, largely in response to perceived inefficiencies among not-for-profit community hospitals, and the idea that existing levels of reimbursement would lead to profit if hospitals were better organized and managed.

In the USA, government hospitals comprise only a small share of hospital beds. These are hospitals that are owned by local governments

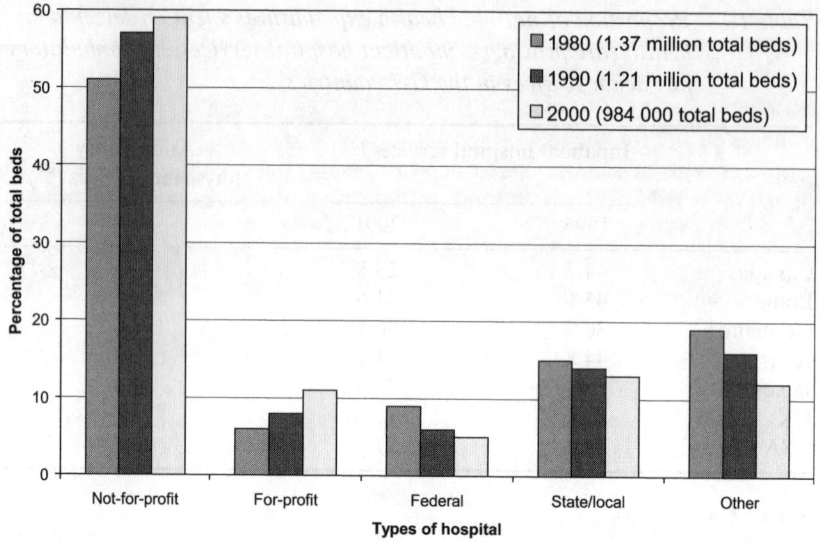

Source: NCHS (2001).

Figure 1.3 Ownership of hospital beds in the USA, 1980–2000

(usually counties) and take on the obligation to serve poor patients who are unable to pay for their care. A small number of hospitals are owned by the national government and care for particular groups of patients, such as veterans, or the chronically or mentally ill. Their number of beds fell by 60 per cent, from 123 000 (9 per cent of the total) to 49 000 (5 per cent).

Other inpatient providers

Providers of inpatient services other than hospitals include nursing homes and assisted residence communities, generally used by the elderly, while rehabilitation facilities offering assistance following acute illness or injury, such as stroke, sports injuries or addiction problems are used by a broader age spectrum of patients.

Ambulatory physicians

The next largest group of health care providers, in terms of expenditure share, is physicians in a non-hospital setting. These physicians can be organized as solo practitioners or in groups as large as hundreds. Some groups are collections of physicians in a single medical specialism, horizontally integrated. Or they may comprise physicians in different, complementary specialisms, vertically integrated into a comprehensive clinic. In the USA, many physician groups, especially smaller ones, are simple partnerships, but larger groups are often corporations, and some of the corporations are actually groupings of partnerships. These large groups are important components in the managed care system, because it is often these large conglomerates of physicians with which third-party payers contract in order to offer comprehensive care to large numbers of enrollees covering a large geographic area.

Other ambulatory providers

Other health service providers practise in diverse ambulatory settings, including pharmacies, laboratories, radiological imaging centres, outpatient surgery centres, patient homes and dialysis centres. As medical technology has changed, new opportunities for reorganization of providers have occurred. An example is the outpatient surgery centre. Formerly, surgery was done only in a hospital, because of the need for stand-by capability for emergencies and the tendency to keep surgery patients in the hospital for many days, both before and after the operation. New forms of surgery are much less risky, however, and so the need for contingency hospital services is often reduced. And, as efforts to reduce expenditures for hospitals have became more intense, it has become apparent that preoperative services could be performed well on an ambulatory basis, and that the recuperation itself could be substantially shortened, often to a matter of hours, rather than days. It has therefore become practical, and economic, to do many kinds of surgery in a free-standing facility without incurring the cost and the risk of infectious disease present in a hospital.

The shift in the locus of care from hospitals to ambulatory or home-based care has produced a substantial, though so far unmeasured, cost to patients themselves and to their family members. Home-based care delivered by family members may be superior in some respects to care delivered in a hospital, but the costs, especially indirect, can be large, especially if family members must give up work or other normal activity or if providing care imposes physical or emotional stress.

Financing Organizations

An essential component of every country's health system is its mechanism that pays for care. In countries with comprehensive single payers, the financing organization is often a part of the ministry of health or finance. Or it may be a private quasi-governmental agency, such as a sickness or social security fund. In the USA, governmental insurance funds are both federal and state. The Federal government administers two of the three largest public insurance programmes, Medicare for the elderly and the Veterans Administration for military veterans. States are responsible for implementing Medicaid, the health insurance entitlement programme for the poor. But the largest share of health insurance funds is controlled by private insurers, whose coverage is a voluntary contract between an insurer and an employer (or, less commonly, an individual). There are hundreds of private insurers in the USA, nearly all of which are publicly owned corporations. Increasingly they are consolidating into multi-state operations in order to achieve economies of scale.

Insurers and health services in all industrialized countries are becoming more and more 'prudent consumers' of care, as they choose 'preferred' providers, negotiate fee schedules and even determine what services are allowable for reimbursement. In the USA, private insurers, and especially managed care organizations, have also become contractors for the government Medicare and Medicaid programmes. Both the federal and state government health agencies have recognized that it is more efficient to enrol publicly insured individuals into managed care plans, so that the government programme pays a negotiated premium and all care is organized and delivered by those plans. Thus managed care is now providing care for both publicly and privately insured individuals.

Manufacturers of Health Care Products

The third component of the health care industry is the set of firms that manufacture health care products, including pharmaceuticals, medical equipment and supplies. Often outputs such as pharmaceuticals are purchased directly by patients. In other cases products are sold by producers to other providers (hospitals, ambulatory physicians, dentists, clinics and so on). For pharmaceuticals, the purchasing decision is especially complex, because the physician chooses the product on behalf of the patient, and the patient purchases it from a pharmacist (who may have additional decision-making authority, especially when the product is produced by multiple manufacturers. Though the patient consumes the product, the third-party payer (public or private) pays for it (see Schweitzer, 1997).

Pharmaceuticals

The shares of pharmaceuticals in the G-7 countries' health budgets increased in most cases from 1990 to 2000, apart from Japan, which registered a remarkable decline from 1995 to 2000. The share in 2000 varied widely from a low of 11.9 per cent in the USA to a high of 22.2 per cent in Italy, as shown in Table 1.8. The share of health budgets allocated to pharmaceuticals must be looked at carefully because it is a ratio of

Table 1.8 Proportion of spending on pharmaceuticals in the total health expenditures in G-7 countries

	1990	1995	2000
Canada	11.5	13.8	15.7
France	16.9	17.6	20.4
Germany	14.3	12.7	13.6
Italy	21.2	20.9	22.2
Japan	21.4	21.6	15.9
United Kingdom	13.5	15.3	n.a.
United States	9.2	8.9	11.9

Source: OECD (2003).

expenditures on pharmaceuticals to expenditures on all health services. A low ratio does not necessarily mean that pharmaceutical expenditures are low, because it might signify only that aggregate health expenditures are unusually high.

Furthermore pharmaceutical expenditures are the product of two variables, pharmaceutical prices and quantities. Therefore expenditures can be high because either prices are unusually high or consumption is unusually high. Lastly, both price and quantity are often influenced by government policies in respective countries. In Japan, for example, at the time of these data, physicians earned a substantial proportion of their incomes by dispensing (selling) pharmaceuticals, a practice that is not permitted in the USA. France and Italy have had especially strong controls on drug prices, which kept prices relatively low, while allowing quantities to rise.

Medical equipment

Medical equipment is typically used in the provision of health services, but can also be used elsewhere in the manufacturing sector. For example, the human genome project would have been delayed for years, but for new devices that permitted rapid assays of biochemical reactions. Just as medical science has grown exponentially, so has the pace of supportive technology. This is in large part the result of increasing specialization within medical fields. First x-ray images of the head were replaced by CT scans. Then CT was replaced for some patients by the MRI. And now neurologists are demanding PET (positron emission tomography) scans, a specialized imaging process of brain function.

Similarly, minimally-invasive surgery is replacing traditional surgical techniques for an ever wider range of applications, but each application requires its own set of instruments and training (for example, gall bladder and coronary artery bypass).

While much of medical technology is utilized strictly for patient care, and was developed for this purpose, some technologies are spin-offs of other technologies. Film-less radiology is a development of clinical radiology and computer imaging. Telemedicine is an enhancement of teleconferencing and other computer and telephony technologies. It is clear that technological development in medical equipment will continue to grow along with growth and specialization in medical science and other 'new' industries.

The health care manufacturing industry, including pharmaceuticals, biotechnology, diagnostic equipment and medical devices, is remarkable for its dependence on research and development investments. Table 1.9 shows the level of R&D expenditure in one of these industries: pharmaceuticals in selected countries. In the USA the rate of investment in R&D exceeds that of any other industry, including electronics and aerospace.

Investment in innovation has been successful in bringing to market large numbers of new drug and equipment products. Some of these are truly innovative, representing breakthroughs in biological science. In some cases these new products bring about improved treatment for conditions, and in other cases technology enables treatment of problems that previously had been untreatable. Other products are more duplicative of existing products in the market, such as competing brands or generic drugs, but their market entry tends to lower prices and gives patients more choice among products that offer slightly different efficacy, side-effects or convenience (see Comanor, 1986). Goldsmith (1999) suggests that new technology, especially biotechnology, offers the potential of eventually reducing overall health costs.

Table 1.9 Spending on pharmaceutical R&D in selected countries, 1998 (€ millions)

	R&D spending
France	2433
Germany	2692
Italy	762
Japan	4426
United Kingdom	3396
EU	12034
United States	15022

Source: Farmindustria (2000).

But innovation comes at a high price. First, the price of new drugs and equipment is frequently substantially higher than that of products that are replaced. Though R&D expenditures do not determine short-run prices, they obviously have to be covered in the long run, and so one finds that prices of drugs that provide new benefits to patients tend to reflect high research and development costs. High prices of new drugs and equipment often raise costs of therapy. These cost increases may be offset by improvements in quality, but, whether or not quality rises sufficiently, either patients or their sickness funds or insurers must decide which therapies to cover.

Innovation also requires that physicians and other decision makers must be informed about new products. The traditional approach of drug company marketing relies upon sales people promoting the merits of products made by the particular company. This approach is inefficient, with each sales person representing only a few drugs within a single drug class. Comparative information about alternative therapies, however, is a public good and will not be provided by individual drug firms. A combination of increased continuing education by physicians and government or medical society-sponsored newsletters would improve the quality of prescribing practice, and is especially important in a rapidly evolving marketplace of health technology.

Another implication of rapid introduction of new medical products is the question of patient information and involvement in clinical decisions. The old model of the dominant physician and the submissive patient is clearly inappropriate today. Patients seek numerous sources of information on medical care, including the print and video news media, the Internet, self-help groups and books. Considerable concern is now heard with respect to direct-to-consumer (DTC) advertising in the USA, which has grown rapidly. Though some advocate restrictions on DTC advertising, a wiser approach would undoubtedly be to offer better information to which consumers could gain access and that they know would be more informative than advertisements.

IMPLICATIONS OF THE MODELS

Health Industry Development Policy

The policy implications of the Health Industry Model (HIM) are strikingly different from those of the traditional Health Model (HM). The policy implications of the models have one point in common: production of an acceptable level of health produced within a budget constraint is still a policy objective in both views of the health system. However, the Health Industry Model identifies additional objectives, including the following:

- growth in domestic health goods manufacturing and services;
- export growth in health goods and services;
- the development of competitive R&D in high value-added and knowledge-intensive sectors;
- the development of innovative sectors such as biotechnology, bioengineering and software industries;
- the production of an optimal amount of spillovers between scientific areas.

Health as a Growing Industry

Both models recognize that health is a growing sector. But while the HM only recognizes the need for health systems to adapt to rising demand in compliance with budgetary constraints, the HIM recognizes that a high demand for health care goods and services represents an opportunity for the development of the economies in industrialized countries.

Demand for health services

Health demand is growing in all countries, for a number of reasons. The first is the aging of populations. The aging of populations in developed countries reflects the success of medical technologies that have enabled people to live longer and healthier lives than ever before. Table 1.10 shows the projected increase in the proportion of the population aged 65 and older in each of six industrialized countries to the year 2050. The proportion is

Table 1.10 Projected proportion of older people in the total population, 1996–2050 (per cent)

	Proportion of population aged 65+	
	1996	2050
Canada	12	24
France	15	24
Italy	17	34
Japan	15	30
United Kingdom	15	23
United States	13	21

Source: Hviding and Mérette (1998).

expected to rise substantially for all of the countries, but the increase is particularly dramatic for Italy, where the population share of the elderly is projected to be 34 per cent. This represents a doubling of the share from the 1996 level. Even in Canada and the USA, where the proportion of elderly had been only 12 and 13 per cent, respectively, in 1996, the share is projected to grow to over 20 per cent by 2050.

Figures 1.4 and 1.5 show recent increases in numbers of elderly (rather than proportions) among the G-7 countries. The actual numbers

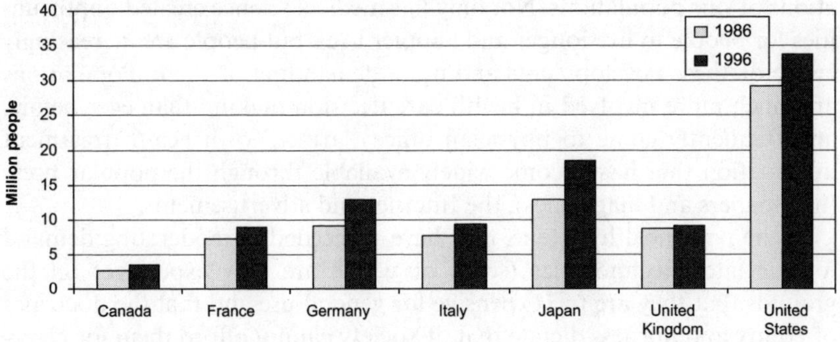

Source: OECD (1998).

Figure 1.4 The rising number of the elderly (aged 65+) in the G-7, 1986–96

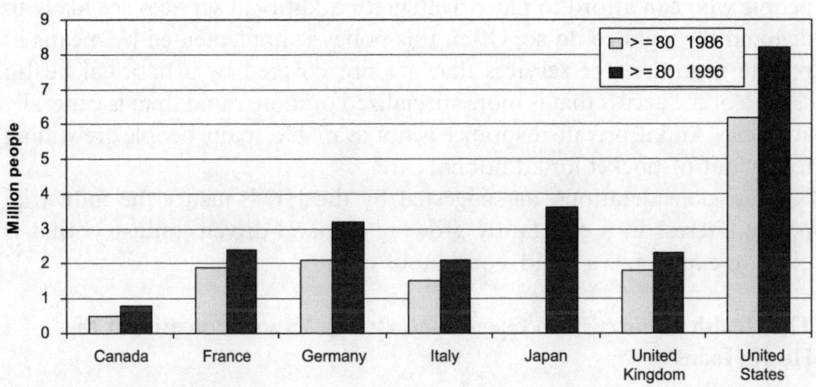

Source: OECD (1998).

Figure 1.5 The rising number of very elderly (aged 80+) in the G-7, 1986–96

better represent the social burden of the elderly population, and are not influenced by relatively high birth rates in some countries, such as the United States. The figures show dramatic increases in the numbers of elderly people in all of the G-7 countries, though the increases for the populations aged 65+ in Germany and Japan are especially dramatic, as shown in Figure 1.4. Figure 1.5 shows the number of people aged 80 and older in the respective years. The proportionate increase in the numbers of these 'very old' people has been especially great in Germany and the USA.

Another factor raising the demand for health services is the rising expectations of our populations. Not only has medical science created opportunities for people to live longer and happier lives, but people are increasingly aware of these developments and more demanding of them. Populations are much more involved in health care decision making than ever before, and frequently come to physician offices 'armed' with health treatment information that has become widely available through the popular press (newspapers and magazines), the Internet and advertisements.

In the past, health systems may have succeeded in moderating demand for the latest technologies (some of which are very expensive) on the grounds that they are too expensive for general use and that the doctrines of equity and fairness dictate that, if society cannot afford them for *everyone*, then these technologies should not be available to *anyone*. Today people are more likely to demand the best quality of care that is available. However, it is unlikely that health systems will be funded to a level that allows provision of the latest technologies to everyone. Under these circumstances, a general standard of access is set for the population, but people who can afford to pay privately for additional services are likely to demand the right to do so. Often this policy is implemented by means of private insurance for services that are not covered by a national health service or for access that is more specialized or more rapid than is generally available. And if private insurance is not available, many people are willing to pay out-of-pocket for additional care.

These considerations, as suggested by the HIM, justify the industrial policy interest in a constantly growing demand-driven 'industry' that is likely to experience a rapid, continuous rise.

The Health Industry in an Open Context: The Export Potential of the Health Industry

The Health Industry Model highlights three additional issues for the health industry. Even if domestic health demand is not growing sufficiently to stimulate research and development, domestic demand is not directed enough towards innovative industries, or a country's health system is not efficient enough to create incentives for innovation, worldwide demand now offers opportunities for the health industry to expand and innovate, transforming a strictly domestic industry into an export industry, in an ever more open and global environment.

National health systems are increasingly open to international competition. The first health market to exhibit international trade was undoubtedly represented by the flow of patients across international frontiers to seek medical care. A lingering barrier to the international flow of patients is financial, as health systems must decide whether to cover services in other countries as if they were performed in the 'home' country. Obviously this is difficult where relative prices differ greatly, and this is likely to be the case between countries that are wealthy or those with weakly administered health systems (and, hence, high costs) such as the USA and poorer countries or those with highly regulated (and, hence, low-cost) health systems.

Though patients are generally treated in their own country, international trade occurs in some other medical services. Increased commerce in technical services (such as laboratory analyses) occurs routinely between the USA and Canada, and among some European countries. Communications technology has created telemedicine, enabling consultations with experts through conference calling or transmission of real-time clinical information, such as electrocardiogram or radiographic data. This technology allows access to the very best specialists, regardless of where the patient is. The technology also allows the storage of records electronically so that old images and records can easily be retrieved from archives. And of course the Internet enables the free flow of information in ways never used before.

Along with the health services market, an international trade of health care equipment and supplies, such as medical equipment and pharmaceuticals, has also developed. This is true for a few countries involved in the trade of laboratory and radiology equipment and pharmaceutical and biotechnology technologies. The health industry trade has recently been further developed on a global scale by the Internet, as in the case of many other industries that benefit from the e-commerce. An interesting example is the 'grey market' for pharmaceuticals between the USA and Canada. Because Canada aggressively negotiates drug price discounts from manufacturers and the USA does not, Canadian prices are often considerably lower than they are in the USA. The number of Americans purchasing Canadian drugs over the Internet has grown rapidly in the last five years. The growth has prompted multinational drug companies to attempt to restrict the flow of drugs into Canada in an effort to reduce re-exports to the USA in order to maintain sales at high prices in the USA.

Spillover Effects of the Health Industry

Many of the health industries, especially pharmaceuticals and medical equipment manufacturing, have interconnecting scientific bases. The CT scan, for example, was developed through developments in radiology, computer science and mathematics (responsible for creating the algorithm that translates axial x-ray readings into two-dimensional images). New developments in genome research are based on biochemistry and genetics. It is becoming clear that new approaches to the treatment of many of today's most serious diseases, such as cancer and AIDS, will not emanate from a single scientific field, but will result from simultaneous developments in genetics, cellular biology and biochemistry. This implies that research in one discipline may have implications for knowledge in others that are more widespread than were thought when the research was begun. R&D is therefore an enterprise much like a multi-product firm, and evaluating its results along a single dimension is likely to understate the value of its output.

Whether research is undertaken by the private sector, by universities, or by government, one must consider the effects of spillovers to other health industry components and, more generally, to and from other industrial sectors. Therefore investments in health-related technology are likely to have implications in fields as diverse as computer science, material science and telecommunications. And the results of successful innovation are likely to be felt widely, beyond the sector where they were originally generated.

A greater capability must be developed for analysis and understanding, and for an industrial policy to be implemented by all countries competing in international markets.

Market-oriented Mechanisms to Increase Efficiency of Health Supply

More countries are adopting health system reforms that will increase provider efficiency. Many of these reforms involve a removal of financing and control to more local units of administration: health regions, areas or districts.

Other reforms create market-like incentives to encourage greater efficiency or the production of services that are more sensitive to patient demand. For example, France, Italy, the USA and the UK have all instituted programmes that reimburse providers according to actual output (measured by case-adjusted volume of service).

This increased efficiency will have many effects. The first will be to increase demand, as consumers benefit from higher-quality, more efficiently produced services. Increased demand for consumer services will also increase demand for intermediate products such as medical equipment. Lastly, improved supply efficiency will increase exports of health goods and services.

According to the traditional Health Model approach, health system efficiency is a means to increase the level of health status achieved within the constrained budget. The Health Industry Model, however, sees that improvement in national health system efficiency also from the point of view that this increased efficiency will also increase demand for intermediate products such as medical equipment. Increased demand for health services goods and services can attract capital to the sector, promote innovation and technological progress and make R&D investments more profitable. Lastly, in a more open context, a national health industry led by more efficient health service supply will increase exports of health goods and services.

CONCLUSIONS

Over the past three decades, the focus of policy debate on the 'health sector' in almost all industrialized countries is on one objective: how to produce an acceptable level of health for the population within a constrained budget (the Health Model). In the Health Industry Model, the focus of the analysis is on the broader potentialities that the health industry can supply to advanced economies. This approach views the health sector as a broad industrial grouping including all the industries involved in one way or another in health care-related activities.

From a political point of view, the HIM adopts the industrial policy approach and identifies other policy-related objectives. These include, for example, the development of competitive R&D in high value-added and knowledge-intensive sectors, development of innovative sectors such as biotechnology, bioengineering and telemedicine and the production of an optimal amount of spillovers.

Both models accept the need to accommodate the increased demand for health. But, if the traditional HM focus is on the need to adjust health systems to this rise in demand within a budget constraint, the HIM recognizes that a high demand in health care goods and services must also be seen as an opportunity for the development of the economy in industrialized countries, in terms of growth, employment and competitiveness in international markets.

NOTES

1. The terms 'entitlements' and 'capabilities' are used by many authors to refer to different concepts. Besides Sen's contributions, see also Dahrendorf (1990), North (1990), Nozick (1974) and Rawls (1971, 1993).
2. This behaviour also applies to private health insurance plans in the USA, where governmental budget pressures are replaced as the constraining force by competitive forces in the health insurance market.

REFERENCES

Anderson, R., 'Revisiting the Behavioral Model and Access to Medical Care: Does it Matter?', *Journal of Health and Social Behavior*, **XXXVI**(1), 1995, 1–10.

Arrow, K.J., Cropper, M.L., Eads, G.C., Hahn, R.W., Lave, L.B., Noll, R.G., Portney, P.R., Russell, M., Schmalensee, R. and Smith, V.K., 'Is there a Role for Benefit–Cost Analysis in Environmental, Health, and Safety Regulation?', *Science*, **272**(5259), 12 April 1996, 221–2.

Comanor, W.S., 'The Political Economy of the Pharmaceutical Industry', *Journal of Economic Literature*, **XXIV**, September 1986.

Dahrendorf, R., *The Modern Social Conflict. An Essay on the Politics of Liberty*, New York: Weiden & Nicolson, Bari: Laterza, 1990.

Dirindin, N. and Vineis, P., *Elementi di economia sanitaria*, Bologna: Il Mulino, 1999.

Enthoven, A.C. and Singer, S.J., 'Markets and Collective Action in Regulating Managed Care', *Health Affairs*, **XVI**(6), November/December, 1997, 26–32.

Evans, R. and Roos, N.P., 'What is Right about the Canadian Health Care System?', *Milbank Quarterly*, **LXXVII**(3), 1999, 393–9.

Farmindustria, *Indicatori Farmacentici*, Rome, 2000.

Goldsmith, J., 'The Impact of New Technology on Health Costs', *Health Affairs*, May/June 1999.

Hviding, K. and Mérette, M., 'Macroeconomic Effects of Pension Reforms in the Context of Ageing Populations: Overlapping Generations Model Simulations of Seven OECD Countries', OECD Economics Department Working Paper ECO/WKP(98)14, OECD, Paris, 1998.

NCHS (National Center for Health Statistics), *Health United States*, Department of Health and Human Services, 2001.

North, D.C., *Institutions, Institutional Change and Economic Performance*, Cambridge: Cambridge University Press, 1990.

Nozick, R., *Anarchy, State and Utopia*, New York: Basic Books, 1974.

OECD, *Health Data 98* (cd ROM), Paris, 1998.

OECD, *Health Data 99* (www.oecd.org/els/health/ecosantefad.htm), Paris, 1999.

OECD, *Health Data 2002* (www.oecd.org), Paris, 2003.

Rawls, J., *A Theory of Justice*, Cambridge, MA: Belknapp Press, Harvard University Press, 1971.

Rawls, J., *Political Liberalism*, New York: Columbia University Press, 1993.

Rice, T., 'Can Markets Give Us the Health Care System We Want?', *Journal of Health Politics, Policy, and Law*, **XXII**(2), April 1997, 383–426.

Rice, T., *The Economics of Health Reconsidered*, Chicago: Health Administration Press, 1998.
Schweitzer, S.O., *Pharmaceutical Economics and Policy*, New York: Oxford University Press, 1997.
Sen, A., 'Economics and Health', *Lancet*, **354**(Suppl. SIV20), December 1999a.
Sen, A., 'Health in Development', *Bulletin of the World Health Organization*, **LXXII**(8), 1999b, 619–23.
UNDP, *Human Development Report 1999*, New York, 1999.

PART TWO

The Macro View

2. Healthy governance: economic policy and the Health Industry Model

J. Robert Branston, Lauretta Rubini, Roger Sugden and James R. Wilson

1 INTRODUCTION

Health care is often regarded as a priority, as suggested by the resources allocated to it in many countries. The proportion of GDP spent on health in 2001 was 13.9 per cent in the USA, for example, 10.7 per cent in Germany, 9.5 per cent in France and 9.7 per cent in Canada (OECD, 2003). However, it is not always clear why health care really matters and, within health care, what determines its impact. One perspective might stress that it is essential to life: without a sufficiently high level of health, individuals cannot fulfil their potential as people. Another view might be, for example, that economic production requires a supply of healthy workers who can carry out appropriately assigned tasks. Or perhaps the health sector is important because it is a source of high-quality employment or of high-technology innovation, the type of performance concerns that the Health Industry Model (HIM) might emphasize. These and related approaches have different implications for resource allocation, for amounts allocated to health provision and for the types of health care that are prioritized. Moreover analysis of the processes by which societies set their priorities raises the possibility that the wishes of entire communities are not being served in particular situations.

Adopting a perspective suggested by consideration of the Health Industry Model, the aim of this chapter is to discuss the importance of the health industry and to analyse the determinants of its impact. We emphasize the democratic governance of the organizations and institutions that influence and comprise the industry, suggesting that this is a fundamental determinant of economic success. One objective is to introduce policy issues. Another is to identify the beginnings of a research agenda that follows from our perspective.

The chapter is organized as follows. Section 2 provides the foundations for discussion and analysis by briefly considering the ambit of the industry. It follows very closely the HIM as advocated by Di Tommaso and Schweitzer in Chapter 1 of this volume. Sections 3 and 4 offer our perspective, drawing heavily on Cowling and Sugden (1998, 1999) and Sugden and Wilson (2002). Section 3 contemplates why the health industry matters and section 4 looks at the determinants of its impact. Based on those considerations, section 5 indicates some items for a future research agenda, focusing in particular on topics that are explicitly relevant to public policy. Our concluding comments are presented in section 6.

2 THE HEALTH INDUSTRY DEFINED

Most existing ways of looking at health care tend to focus their debate on how to achieve an 'acceptable level' of health for the population in question, given constrained budgets, and thus the need for choices to ration resources between all of the available options. Having reviewed this so-called Health Model (HM), Di Tommaso and Schweitzer argue in Chapter 1 of this volume that recent 'trends towards high rates of research and development in the industry, more sophisticated and advanced technologies, use of highly trained personnel and the opening of international boundaries to the flow of health care services suggest the need for a new analysis'. They propose the 'Health Industry Model', which groups 'related health care activities as one of the major "industries" in advanced economies' and thus focuses on the 'broader potentials that the health industry can supply to advanced economies'. In other words, they view the health industry as a broad industrial grouping.

Di Tommaso and Schweitzer identify three main parts to the industry: the providers of health care, financing organizations and the manufacturers of health care products. These wide categories include organizations such as hospitals, other inpatient providers (for example, nursing homes or rehabilitation facilities), health insurers (both public and private), pharmaceutical and biotechnology companies, as well as medical equipment and other suppliers.

We would suggest that there is a case to be made for adding a fourth main part: the providers of learning (education and training) in health care. Whether the split is into three or four categories, however, the basic point is that the health industry consists of all the corporations and other organizations/institutions that are in some way involved with health care-related activities. It is from this basis that we now consider the industry's importance to economies.

3 WHY DOES THE HEALTH INDUSTRY MATTER?

Three Reasons

An implication stressed by Di Tommaso and Schweitzer in the HIM is that health matters because, viewed from a certain perspective, it is a sector with potentially significant impacts on key economic indicators. These include wealth creation, employment, international trade, innovation and, seen to be especially significant, the development of high-technology industries. The most important original element introduced by Di Tommaso and Schweitzer is probably to recognize that health is in a sense like any other industrial sector, and consequently that it needs analysing as such. Their insight is therefore to highlight the fact that its impact extends significantly beyond that identified in traditional approaches to health care, and that health is in fact an especially important industry.

Notwithstanding, and indeed taking on board, that insight, we would argue that in fact the health industry matters to the development of economies for three reasons:

1. because the health industry has effects on attainment of the aims and objectives of people through its direct impact on aspects of performance – the development of high-technology activity, employment and so on;
2. because people might see their health and especially particular aspects of it as important to them;
3. because health is a key requirement for successful economic development founded upon democratic inclusion.

Although it is the first of these that is stressed by Di Tommaso and Schweitzer, the others are also related to their insight. Traditional analysis of health according to what they term the Health Model has possibly helped to keep it apart from the study of industrial economic development and of associated economic policy; health seems to be widely neglected by academics dealing with development, and it is rarely included as a central issue in manuals of development economics. However, once health is recognized as an industry, this distance is reduced. Consequently the question 'why does the health industry matter?' can be answered through a consideration of the wider literature on the development of economies and on associated public policies. It is that analysis that leads to the second and third reasons, above. Moreover that analysis also provides insight into the exact meaning of the first, for example by providing an underlying rationale for the significance (or otherwise) of high-technology industries.

The Development of Economies

We suggest that analysis of the actual and potential role of the health industry depends on the conceptualization of 'the development of economies' (whether those economies are labelled 'developed', 'less developed' or 'developing'). Typically development is framed in a relatively narrow sense, in terms of performance measured by a specific set of indicators (including health, education and income). Moreover these indicators are frequently determined externally, where they are strongly influenced by institutions such as the World Bank and IMF. They are also associated with a so-called 'free-market perspective' and correspondingly with market-centred policies. Insofar as they have an economic rationale, such policies are rooted in the market failure approach promulgated by mainstream economists.

Following Sugden and Wilson (2002), a fundamental problem with such an approach is that it may not reflect the aims of those that are seeking to develop. This is incongruous, because a tenet of the market system underpinning these development attempts is that others are not best placed to decide what is appropriate for an individual or organization; for example, government is generally not best placed to decide what is optimal for a firm. Furthermore external prescriptions run the danger of stifling the ideas and processes that stem naturally from people within a particular territory or community, and that might offer innovative and contextual solutions to the development problems that those people face.

An alternative approach is to conceptualize and define development specifically in terms of the aims of those seeking to develop (which, of course, is not to preclude cooperation with external sources). To illustrate, consider someone who lives in a locality in Italy, in the city of Ferrara, for example. What does the term 'economic development' mean to that person? Perhaps it means a preference for higher standards of health care and/or education for children. It might mean the opportunity to earn a higher income than at present, perhaps through the creation of productive, high-technology industries. Maybe the desire is to create more or better-quality jobs, possibly to alleviate social problems in the city. 'Economic development' might also incorporate a wish to build on the experiences, traditions and cultures of the community. Perhaps there is an attachment to certain activities and ways of doing things, and maybe a perception of enhanced status through being involved in modern, fashionable and/or high-technology industries. Perhaps the community has suffered a particular health scare, such as the uncommonly frequent incidence of a specific disease, making it a priority in the economic development of that community to address that disease. However, while we can 'guess' at the meaning of economic development for Ferrarese, as 'outsiders' we cannot hope to be totally accurate.

Compare with this the meaning of development in Ferrara for those who make key decisions affecting its future. The city is part of the region of Emilia Romagna and the nation of Italy. At regional and national levels motivations and desired outcomes may differ from those found in Ferrara. For example, policies formulated for Emilia Romagna in Bologna or for Italy in Rome might reflect the interests of groups or individuals that are particularly successful at lobbying. Moreover, given the physical and cultural distance between people in Ferrara and the decision-making apparatus in Rome, attaining transparency and accountability in such decisions is likely to be problematic. Similarly, important decisions are made outside Italy, for example in the European Union and by foreign-controlled corporations. Their aims are likely to differ from those of the Ferrarese (as implied by Bailey et al., 1998, discussing the aims of foreign direct investors in Central and Eastern Europe).

Why does the development of an economy or of a sector within that economy matter? It depends on who is answering, but in essence 'development' is only meaningful to the people seeking to develop when that development reflects their own aims and objectives (Sugden and Wilson, 2002). We therefore suggest it is this that provides the criteria for identifying why the health industry matters.

Accordingly the health industry is important to the development of an economy for three reasons. First, it matters to the extent that the people in that economy have aims and objectives (in terms of, for example, wealth creation, employment, international trade, innovation and the development of high-technology industries) that are affected by the health industry as a producer, employer, trader, innovator and so on. Second, it is significant to the extent that people in that economy see people's health as an aim in its own right. Moreover the basis of this entire argument is the inclusion of all people in the democratic identification of the aims and objectives of economic activity. It follows that the third reason that health matters is that people's levels of health influence the possibilities for, and the potential of, democratic participation. For example, maximizing the potential gain from people's inclusion in decision making necessitates that from an early age they have (amongst other things) the nutrition that enables them to develop fully their intellectual abilities.

4 WITHIN THE HEALTH INDUSTRY, WHAT DETERMINES ITS IMPACT?

The premise that all people in an economy need to have strategic input into the determination of their own development paths follows from and highlights the importance of decision making and democracy. The consequence is that, in designing the structure of an industry and its components, it is necessary to consider channels for people to participate in strategic decision making in the sector as a whole. Such decisions might then feed into, and become integrated with, democratic decisions concerning development more generally. This is also in line with the more general argument that the crucial determinants of an industry's impact are its strategic decisions.

Following Cowling and Sugden (1998, 1999), the making of strategic decisions is central to understanding the impact of economic activity. They are the decisions that give rise to the broad plans of corporations and other organizations/institutions. These plans are the crucial determinants of the activities with which organizations and institutions are associated. Such an approach has its origins in the seminal contribution to the theory of the firm by Coase (1937), suggesting that 'firm' and 'market' are alternative means of coordinating production and arguing that a type of planning occurs within a firm that is associated with its bypassing of markets. The importance of strategy is identified in Zeitlin (1974), integrating intra-firm decision making with the wider debate on corporate governance. He saw the ability to control as the ability to determine a corporation's broad, guiding policies. These policies can be equated with strategic decisions and therefore, as Zeitlin saw it, to control these is in essence to control the corporation.

Adopting the assumption of standard economic theory that individuals make decisions in their own best interests, it becomes crucial to identify who makes strategic decisions. For the case of modern corporations, for example, this is a subject of considerable debate. However, one thing is agreed: control is exclusive, resting with a sub-set of those who have a vested interest (Branston et al., 2003). This consensus is vital because, by their very nature, strategic decisions are the most important determinants of the effects of a corporation; different decisions will result in differing impacts on the economies where a corporation operates.

Illustrations of strategy and of variation in interests in the health industry can be readily provided. They include, for example, choices over whether or not to develop major new products and processes, or to undertake large-scale and innovative high-technology investments. Views on such decisions are likely to differ across health industry administrators, managers, employees and consumers, for instance. In addition, the work by

Zeitlin (1974) on corporations points to organizations making strategic decisions over relationships with other organizations, with governments and with employees.

These encompass, for example, the decision of GlaxoSmithKline to offer to supply two of its antibiotics to the US government free of charge, in order to help combat the threat posed by anthrax in that country.[1] In issue here were corporate–government and possibly corporate–corporate relations, bearing in mind that the rival firm Bayer believed its interests were best served by continuing to sell its anthrax antibiotic, Cipro, in the conventional way.[2] As another example, strategic decisions over relationships with employees are illustrated by the case of Anglo American, a South African mining company that has opted to provide employees with AIDS treatment at the company's expense. Anglo American estimated that 23 per cent of its workforce in Southern Africa was HIV-positive, adding US$5 to US$6 an ounce to its cost of gold production, US$3 an ounce to its cost of platinum production and, in the words of a company representative, the decision to provide treatment was expected to yield benefits 'through extending the lives of infected employees'.[3]

Beyond corporations there are similar strategic concerns. The (non-corporate) health systems of many countries, for example, provide medical drugs to patients either totally free or at (flat rate) subsidized tariffs. However, not all drugs are included in such programmes and so decisions must be made as to which to include, and which to exclude. These are questions of strategy. A specific illustration is given by access to the anti-impotence drug, Viagra, in the UK. It was initially blocked from being prescribed, except in exceptional circumstances, owing to fears that excessive demand could cost the government-funded health service more than US$1.5bn.[4] Guidelines were later issued which decided that the drug would only be available to certain groups, representing approximately 17 per cent of potential beneficiaries, and even then only in limited quantities.[5] Viagra was only the seventh drug in the UK to have had its availability so restricted on cost grounds.[6]

In countries where the university sector is state-funded, the decision as to how many doctor, nurse or other medical training places to support is another example of a strategic decision outside a corporation. The number of staff trained will have a significant impact upon the ability of the health industry as a whole to provide appropriate medical services. Without a continuous supply of new staff, the industry would be unable to fulfil its obligation of providing medical care to local communities. Indeed a similar situation also exists in hospitals where various different types of appointments exist. The number of appointments available, and the balance between medical professionals who are continuing their education and those who have special skills to teach, again determines how the industry is able to provide its services both now and in the future.

To consider variation in interest, the anthrax case is again revealing. People would have a clear interest in access to medicines, both in the event of a specific scare and also more generally in order to allay their fears. This interest contrasts with the aim of the corporate manufacturers: to make profit from their activity. It also compares with the position of governments, which might (at least to some extent) reflect wider concerns. Indeed, in the midst of an anthrax scare, the US and Canadian governments were close to overriding the patent of Bayer, so that greater quantities could be manufactured at lower costs. With potential legal action involved, both countries struck agreements with the corporation over supply and price.[7] Similarly, on other occasions, countries such as South Africa and Brazil have been on the receiving end of criticism (and other more direct action) for measures to allow, or for allowing, the generic production of patented drugs. Cheap access to, for example, treatment for AIDS is vitally important to such countries, but is argued to be against the interests of the pharmaceutical companies whose patents are being ignored, and which are thus losing revenue (Klein, 2002). Historically, these corporations have enjoyed the support of the USA in their efforts to prevent the production of patent-breaking drugs. The change of heart on this strategy by the USA in the midst of its anthrax scare shows not only how interests differ between governments and corporations, but also how events can cause those interests to be revealed in changing strategies.[8]

Variation in interest in strategic decisions can also be illustrated by the election of Richard Taylor to the British Parliament, for the constituency of Wyre Forest. Taylor stood for the Independent Kidderminster Hospital and Health Concern Party, protesting against the downgrading of the hospital in Kidderminster, a town in the English Midlands, and arguing against relocation of its accident and emergency unit to a facility 12 miles away. It appears that Taylor's dissatisfaction with this strategic decision was widely felt in the local community. The incumbent Labour Party (which formed the national government, in turn controlling the hospital in question) saw its 26 843 votes in the previous (1997) election collapse to just 10 857; an earlier Labour majority of 6 946 gave way to a majority of 17 630 for Taylor.[9] A significant number of local people clearly held a view of the strategic decision to downgrade Kidderminster hospital, a view at odds with that of the actual decision makers.

It also ought to be recognized that variation in interests is not always reflected in a precise distinction between easily identifiable groups, as might seem to be the case in examples given above. There may be differences between, for instance, patients seeking a certain sort of medical treatment or corporations producing similar types of pharmaceutical products. Consider the recent fears over the combined child vaccine for measles, mumps and rubella (MMR). Many parents have decided not to have their child immunized with this vaccine over fears of a link with autistic-like disorders and bowel problems. In the UK (in September 2001) approximately 87 per cent of children in the appropriate age bracket were being immunized whilst, in the Republic of Ireland,[10] the Eastern Health Board of Ireland reported rates of only 74 per cent (in January 2001).[11] There were seemingly differences of opinion across parents.

In short, then, this analysis suggests that the prime determinant of the impact of a sector is its governance. Because interests vary and because decision makers act according to their own aims and objectives, it is crucial to understand who makes the strategic decisions in the organizations and institutions that influence and comprise the health industry.

5 POLICY POSSIBILITIES AND FUTURE RESEARCH

Faced with such analysis and evidence for other sectors and for economies in general, it has been argued that economies can be characterized by strategic failure: concentration of strategic decision-making power in the hands of a few implies a failure to govern production in the interests of the community at large (Cowling and Sugden, 1999). This has been seen as a particular problem in both free market and centrally planned economies, for example. To avoid strategic failure, it has been further argued that the democratization of governance is required, to reflect the concerns of everyone interested in production activity. The existence of separate and distinct preferences over strategic decisions implies a clear choice that needs to be recognized in economic analysis and policy. The choice is between seeking to include all parties with a democratic voice in the production process, to enable a strategic direction that serves all of the people, or basing development on the narrower interests of an exclusive group.

Inclusive, democratic strategic choice would require appropriate public policies, as Cowling and Sugden (1998) suggest in the context of different types of corporation. Moreover, it should be recognized that this would be a very different foundation for policy than the market failure approach (as outlined by Di Tommaso and Schweitzer in the Introduction to this volume). Whereas our focus is strategy, market failure is rooted in a market-centred analysis of the nature and impact of firms, and of consequent economic development possibilities.

At one level our conclusion from this discussion is therefore that the structure of the health industry and of its components needs to be in a form that recognizes and accommodates variations in interest, ensuring that exclusive concerns do not dominate and recognizing the need for democratic involvement. For example, policy mechanisms need to be found to widen participation processes.

At another level, however, our analysis has far deeper implications. If development rests on successfully incorporating people's aims and objectives, then democracy in all areas of economic life is central to development. Moreover democracy itself is inherently not mechanistic (see, for example, Dewey, 1916; Hirschman, 1970; Sunstein, 2001).[12] We would emphasize, therefore, that mechanisms should be seen as means and not ends. Their role is to facilitate wider processes of learning to engage, learning to participate and learning what it means to be part of truly democratic governance processes.

Future research is needed to improve understanding of these wider processes. Mechanisms that encourage an inclusive approach in key sectors would allow the evolution of institutions and organizations within those sectors to be brought more into line with the development aspirations of the people that are affected by their activities, whether this be in terms of more high-technology industry or any other aim that people might have. This might enable the beginnings of a more deep-seated change in attitudes to engagement and cooperation that might grow to permeate other arenas in economic life, leading to more democratic development.

Given this concern with the way decisions are made, we suggest that it would be appropriate to investigate decision making within particular parts of the health industry. For example, evidence points to one potentially interesting option being an exploration of the health care sector of England and Wales. Hutton (2000, p. 1) reports on a survey carried out in March 2000, and which indicated that 63 per cent of people thought Britain's National Health Service (NHS) the 'most valuable institution' in the country.[13] He argues that 'the NHS is tasked to provide equal care to every British citizen on the basis of their equal need, irrespective of where they live or how much they earn. The service is publicly owned and accountable, and is almost wholly financed by general taxation' (ibid., p. 1). However only 4 per cent of people thought that the NHS provided a good service that could not be improved upon, whilst 62 per cent believed that it required to be improved 'quite a lot' or 'needs a great deal of improvement' (ibid., p. 3).

A particularly interesting finding is that only 36 per cent of the people surveyed believed that the NHS was run 'in an open way and consults the public' (ibid., p. 126). This points to people viewing strategic decision making within the NHS as being undemocratic, the preserve of a small group. Indeed, 77 per cent of people thought that they had little or no power over their medical treatment, whilst 55 per cent thought that they should have a lot of power in this respect, and only 3 per cent thought that they should have no power at all (ibid., pp. 5, 126–7). These apparent concerns with governance, and the fact that the British government has committed itself to a period of reform and to increasing NHS spending by 6.1 per cent in real terms (ibid., p. 3), make this an interesting case for investigation. Comparisons with other countries might also be especially beneficial.

Investigation of the British case would also run into debates about how health care is funded. This is a topical issue in many countries, and in Britain it has been framed around a dispute over private versus public control. Policy proposals from the British government have envisaged the creation of 'foundation hospitals' accountable to and controlled by local communities; they would have greater room for manoeuvre in raising and spending funds than other hospitals, and would be established as non-profit, public interest companies controlled by a council that includes local patients, residents and hospital staff. Critics have suggested that this would result in a two-tier system, the large and powerful foundation hospitals pulling resources from other facilities.[14] It has also been suggested that the new hospitals would seek business abroad, and that they would be run by self-selecting elites.[15] Presented by the government in a language of increased democracy, an important research question is whether this 'democracy' would be more apparent than real.

At root, this is partly a question of whether components of the health industry ought to be controlled and operated along the lines of major corporations in the private sector. For us, this choice and the associated research question turn on the implications for who governs and in whose interests. The fundamental issue is neither market versus non-market, nor public versus private ownership. Rather it is whether the process and mechanisms of governance, for both the individual production units and for the sector as a whole, seek and allow all interested parties to engage appropriately in the democratic determination of resource allocation and distribution.[16] Our scepticism stems from the view that the typical corporation in a market economy is undoubtedly not democratic in this way (Branston et al., 2003). Moreover to create new organizations that are more accountable to a local community does not result in greater efficiency for an economy if those new organizations have sufficient (market and non-market) power to be able to undermine fundamentally the activities of others. This includes organizations that are established in other communities as rivals in a market or a market-type system. That is a lesson from evidence on the abuse of market power in industrial sectors the world over (and would be an *outcome* akin to having a broadly economically successful North America and Europe, and a deprived Central America and Africa, for example). A key requirement for efficiency is an absence of concentrated strategic decision-making power. A corresponding research question is whether or not existing and potential arrangements are in line with that need.

As well as detailed issues relating to components of the health industry in particular places, however, the research agenda must also appreciate the need to stand back and take overviews of the sector as a whole, across a particular locality and across sets of localities, including those at national levels. In doing so the global dimension must be recognized. Stiglitz (2002, p. 222) argues that 'we cannot go back on globalisation; it is here to stay. The issue is how we can make it work. And if it is to work, there have to be global public institutions to help set the rules'. Specifically on health care, he observes that there are 'global . . . issues like the spread of highly contagious diseases such as AIDS, which respect no boundaries. The World Health Organization has succeeded in eradicating a few diseases, notably river blindness and smallpox, but in many areas of global public health the challenges ahead are enormous. Knowledge itself is an important global public good; the fruits of research can be of benefit to anyone, anywhere, at essentially no additional cost' (ibid., pp. 223–4). He goes on to discuss transparency and accountability in international institutions, effectively throwing down a gauntlet for future research. Consistent with the democratic development of economies, 'globalization' necessitates an understanding of which aspects of health care are best provided locally and which are suitable for multi-locality (perhaps including global) provision. To the extent that global provision appears to be desirable, what processes and mechanisms might ensure that those dimensions of the health industry are nevertheless democratically governed from localities, accountable to and legitimized by those localities? Different forms of globalization imply different processes, mechanisms and structures for the health industry; what are these differences?

6 CONCLUSION

It is clear that there is much that still needs to be learned and understood about the requirements for appropriate governance in the health industry. What is equally certain is that the industry occupies a crucial place in economic analysis, especially when development is based on the need for democratic governance of an economy and of its sectors, organizations and institutions. The issues that we have raised open up new avenues for thought and research. Not only does this agenda need to be expanded, but we also suggest that its pursuit needs to be prioritized.

Can health lead the way in promoting high-technology sectors? Our consideration of the Health Industry Model leads us to suggest various explanations for the importance of health. Corresponding to this we explore what determines its impact and therefore introduce the policies that are required for success. We argue that one reason why it might matter is its impact in terms of high technology, but we see this as *one amongst myriad* possibilities. Accordingly, before suggesting that it lead the way in any particular economy, we really ought to ask the people of that economy whether or not they want it to lead the way. The answer to that question depends upon the preferences of the people. If they answer in the affirmative, however, the way forward according to our analysis is not paved with the market failure approach advocated by many economists. Rather the key is strategic choice. Most especially, it is identifying who makes strategic decisions, and how.

NOTES

1. 'GSK Offers Free Anthrax Antibiotics to America', *The Times*, 27 October 2001.
2. See 'Bayer Beats Canada in Anthrax Row', BBC online news (http://news.bbc.co.uk/hi/english/business/newsid_1615000/1615591.stm) accessed 23 October 2001, and 'American's Anthrax Patent Dilemma', BBC online news (http://news.bbc.co.uk/hi/english/business/newsid_1613000/1613410.stm) accessed 23 October 2001.
3. 'Anglo to Offer Free Aids drugs to Staff', *The Times*, 7 August 2002.
4. http://news.bbc.co.uk/hi/english/special_report/1998/viagra/newsid_237000/237710.stm, accessed 26 June 2001.
5. http://news.bbc.co.uk/hi/english/health/newsid_382000/382682.stm, accessed 26 June 2001.
6. http://news.bbc.co.uk/hi/english/health/newsid_337000/337851.stm, accessed 26 June 2001.
7. See the earlier references on the Bayer example.
8. 'Patent Row Brews Ahead of WTO Summit', BBC online news (http://news.bbc.co.uk/hi/english/business/newsid_1619000/1619690.stm) accessed 26 October 2001.
9. See 'Doctor Wins by Healthy Margin in Wyre Forest', *The Times*, 8 June 2001.
10. See 'MMR Worries "Unjustified"', BBC online news (http://news.bbc.co.uk/hi/english/health/newsid_1556000/1556020.stm) accessed 30 October 2001.
11. See 'Measles: The Irish Experience', BBC online news (http://news.bbc.co.uk/hi/english/health/newsid_1100000/1100174.stm) accessed 30 October 2001.
12. See also Branston et al. (2003), arguing that democracy will not be attained through regulation and law.
13. It is interesting to note that the next highest institution in this poll was Parliament, with only 12 per cent of people voting this the most valuable.
14. See, for example, 'Hospitals Move Prompts Warning of Two-Tier NHS', *The Independent*, 14 November 2002.
15. 'Blair's Hospital plans Under Attack', *The Observer*, 17 November 2002.

16. There is a parallel and related debate in England and Wales regarding, amongst other issues, universities. See, for example, 'Top-up Fees Could Cost New Labour Dear', *The Guardian*, 18 November 2002; and on a university system that mimics and serves transnational corporations, see Sugden (2003). Comparative research on health and education might be a source of useful insight. Indeed the sectors are also related in that education for the health industry is a key issue, our suggested fourth category in a Di Tommaso and Schweitzer type of model (see Chapter 1).

REFERENCES

Bailey, David, Thomas, Rachel and Sugden, Roger (1998), 'Inward investment in Central and Eastern Europe: the compatibility of objectives and the need for an industrial strategy', in Storper, Michael, Thomadakis, Stavros B. and Tsipouri, Lena J. (eds), *Industrial Policy for Latecomers in the Global Economy*, London: Routledge.

Branston, J. Robert, Cowling, Keith and Sugden, Roger (2003), 'Corporate governance and the public interest', mimeo, L'institute, University of Birmingham (www.linstitute.org).

Coase, Ronald H. (1937), 'The nature of the firm', *Economica*, **IV**, 386–405.

Cowling, Keith and Sugden, Roger (1998), 'The essence of the modern corporation: markets, strategic decision-making and the theory of the firm', *The Manchester School*, **66**(1), 59–86.

Cowling, Keith and Sugden, Roger (1999), 'The wealth of localities, regions and nations; developing multinational economies', *New Political Economy*, **4**(3), 361–78.

Dewey, J. (1916), *Democracy and Education*, New York: Macmillan.

Hirschman, Albert O. (1970), *Exit, Voice, and Loyalty. Responses to Decline in Firms, Organizations, and States*, Cambridge, MA: Harvard University Press.

Hutton, Will (2000), *New Life for Health. The Commission on the NHS chaired by Will Hutton*, London: Vintage.

Klein, Naomi (2002), *Fences and Windows. Dispatches from the Front Lines of the Globalization Debate*, New York: Picador.

OECD (2003), *OECD Health Data 2003* (www.oecd.org/health/healtdata), accessed 14 July 2003.

Stiglitz, Joseph E. (2002), *Globalization and its Discontents*, New York: Norton.

Sugden, Roger (2003), 'Internationalism and economic development: transnational corporations, small firm networking and universities', in Waterson, Michael (ed.), *Competition, Monopoly and Corporate Governance: Essays in Honour of Keith Cowling*, Cheltenham, UK and Northampton, MA: Edward Elgar.

Sugden, Roger and Wilson, James R. (2002), 'Development in the shadow of the consensus: a strategic decision-making approach', *Contributions to Political Economy*, **21**, 111–34.

Sunstein, Cass (2001), *Republic.com*, Princeton: Princeton University Press.

Zeitlin, Maurice (1974), 'Corporate ownership and control: the large corporation and the capitalist class', *American Journal of Sociology*, **79**(5), 1073–119.

3. Control, competition and co-operation in European health systems: points of contact between health policy and industrial policy

Giovanna Vicarelli

1 FOREWORD

The health policies that have been adopted in Europe in the last 20 years have been labelled 'restrictive' in order to distinguish them from 'expansive' policies which characterized the previous decades. The increase in health expenditures in Europe in the 1970s, together with the growing concern about the economic sustainability of an expanding sector, drove all European countries to develop specific strategies aimed at reducing and containing both the provision of and the demand for health services.

Given the wide variety of choices, it is difficult to evaluate properly these restrictive policies. As a matter of fact, while a few countries have limited the provision of services, some have increased taxation and some others have decentralized the responsibility for health expenditure. However, in spite of national differences depending on the history of each country's health system, many authors agree on the identification of a certain degree of convergence on policy implementation.

In analysing these forms of convergence three main policies can be identified which correspond to three different approaches for the regularization and social organization of the health sector. The first policy, which is analysed in the third section, consists of a top-down intervention in which the government and public bodies cut down and limit health expenditure, acting at a macroeconomic level. In the second (explained in the fourth section), equilibrium factors are sought at a microeconomic level. In particular an attempt is made to introduce forms of 'managed competition' among health service providers and to distinguish between the production function and the funding function, where no distinction existed. Holland, Sweden and Great Britain were among the countries which followed this policy with greater determination even if similar forms were introduced everywhere in Europe, Italy included.

Finally, the third policy aimed at increasing the coordination and integration capacities of all actors involved in the production and provision of health services within a limited territorial context (managed cooperation – fifth section). The underlying idea is that local authorities are the most competent to manage health expenditure as they know both population needs and current health expenses; therefore they are expected to coordinate and integrate the actions of public bodies and private organizations providing health services.

The relationship which took place between health care policy and industrial policy and their degree of interdependence can be seen for each phase. In particular, while the Health Model (HM) seems to be applicable for the first phase, and the Health Industry Model (HIM), as proposed by Di Tommaso and Schweitzer, seems to apply for the second phase, a third 'Health Industry Integration Model' (HIIM) would rather seem more applicable to the third phase, characterized by higher integration, not only from economic and productive points of view, but also from the point of view of ethical and social values.

This analytical perspective allows us to identify, in the health policies of the 1980s and 1990s, the emergence of new social forms, different from the traditional ones, which, according to Polany's typology, focus on competition, collaboration and control. As shown in the sixth section, managed competition and managed co-operation are intermediate regulatory forms to which, in a six-pole scheme, competition–cooperation can be added.

2 CHANGING STRATEGIES IN HEALTH SYSTEMS

In the history of social welfare, the 1980s and 1990s mark a period of change in health policies almost everywhere in Europe. OECD studies have singled out, in spite of national diversity, a certain degree of convergence of two forms of intervention which characterized this period and which seem to have been complemented more recently by a third strategy. The first type of policy is macroeconomic and aims at controlling health expenditures. The second is of a microeconomic kind, its objective being the increased efficiency of health systems, while the third seems to concentrate on outcomes, although rationalization and control continue to be of primary importance (OECD, 1999, p.404).

According to Pierson (Pierson, 1999), the first is a 'cost containment' policy, which is implemented with the traditional methods of a top-down public policy. Only in a few national cases and in certain fields of intervention can this orientation be seen as a form 'decommodification' owing to the restrictions on access to health services and the reduction in the benefits provided. The second form of intervention, as well as the third, seems rather to fall on the strategy that Pierson calls 'modernization' which implies two forms of change: rationalization involving changes inspired by new ideas on how to pursue traditional objectives, and updating, that is the adaptation of old programmes to meet new social requirements (ibid., p. 404).

What seems to be more evident in the second and third policies is the need to resolve the problems of a top-down regulation incapable of properly governing an open and complex social system. As a matter of fact, given the variety of actors and forms of equilibrium, in a situation of strong interrelation and dependence in which the actors become more and more responsible for choices and results, it is necessary to find bottom-up regulations allowing the system to be dynamic, albeit controlled. According to Bianchi's models, it seems that the choice to intervene with microeconomic instruments should be regarded as a form of 'programming by catalyst concepts', provided in this case by managed competition and the quasi-market, while the third model, centred on the efficacy of the results obtained, is very close to 'programming by integrating setting', that is by settings of integration (Bianchi, 1997).

Therefore, while in a policy of budget restraints the regulatory instrument would be essentially provided by 'direct and indirect controls', as Mossialos and LeGrand demonstrate, in the second model the emphasis seems to be on 'budget setting', that is the capacity of the different actors to manage properly the amount of resources allocated. Finally, in the third case, 'budget shifting' would prevail, aiming at coordinating and possibly integrating the behaviours of the many actors who are responsible for health costs and health care activities. All the policies we have described seem to have been implemented between 1970 and 2000, although the period can be divided into three phases, each corresponding to the dominance of one of the three forms: 'controls during the late 1970s and early 1980s; budget setting in the mid- and late 1980s and budget shifting during the 1990s' (Mossialos and LeGrand, p. 71).

Referring to this conceptual framework, it is possible to say that in Europe attempts at controlling health expenditures have been made with regulations that have differed in time, although the role of the state has never been questioned. This means that the changes of the 1970s–1990s did not question either the state's control of health services or a system of welfare based on solidarity and social redistribution ideals within the health care system. The fostering of services' rationing or their rationalization through competition or co-operation was only a way to enhance the value of the public role and make it economically viable. In this context, however, the continuous development of medical techniques, both diagnostic and therapeutical or preventive, was faced with methods that varied according to the regulatory systems of those periods.

In order to understand these phases better and the different instruments used to contain health expenditures, it might be useful to outline the regulatory principles underlying them, as well as the organizational contexts, which should make them applicable.

3 COST CONTAINMENT THROUGH DIRECT AND INDIRECT CONTROLS

All reforms of health service systems in Europe were adopted in consequence of the increase in health expenditures, which rose from 3.8 per cent of the GDP in 1960 to 8 per cent in 1992.[1] However, the highest increase occurred in the years 1960–80 rather than in the period 1980–97, which saw an average increase of +12.02 per cent rather than 6.69 per cent. Moreover, while in the 1970s hospitalization accounted for the biggest increase, in the 1980s and 1990s the increase was to be attributed to outpatient medical services and the pharmaceutical sector (OECD, 1995). This does not mean that expenditure did not increase considerably in real terms, and sometimes, as happened in Italy,[2] even more markedly than the GDP. It only means that, in the 1980s, expenditure appears to have been more controlled and above all very different from that of the USA, where there was an increase of 8.7 per cent of the GDP (twice as great as the European one) from 1960 to 1992, the increase mainly occurring between 1985 and 1992, with a 3.5 per cent increase in GDP (ibid.).

In this scenario, in addition to real expenditure trends, choices are heavily conditioned by fears for the future, especially a growing demand due to an ageing population, income increase and greater insurance coverage. According to OECD, these factors could cause a remarkable increase in health expenditures, especially if they are combined with technological development, increased staff numbers and health facilities and a real growth in prices. The rising demand for health services should also be put down to growing concern about physical and psychophysical well-being, which the new technologies are able to guarantee. The consequence is that traditional welfare systems, instigated at a time when life expectancy was considerably lower and the culture of health was totally different, are no longer capable of meeting demand that, besides growing, appears to be diversified and more exacting from a technical and relational point of view.

Moreover changes in the funding of health services and in indirect forms of health care provision tend to throw the past systems of welfare off balance. Underlying these changes is, on the one hand, the disappearance of the large occupational base, which supported the fiscal or taxable capacity of the working population and allowed the development of health systems, and, on the other hand, the crisis of family networks which, especially in Latin countries, have often taken the place of deficient health facilities. In other words, the increase in demand and the rising costs due to exogenous factors seem to be offset by less availability of resources and less capacity of the social networks to provide additional or integrative services.

According to this model of analysis, the first instrument used to contain health expenditures is the control of hospital staff numbers (in Ireland and Spain, for example), the price of drugs, hospitalization periods, the number of hospital beds, the amount of capital invested and the purchase of new technologies, medical prescriptions and the possibility of enhancing service provision outside the hospital (for example in Northern Europe). Another measure could be the non-reimbursement for a few categories of pharmaceutical products or the exclusion of some services from health insurance coverage; instead users are asked to contribute to health expenses while health contributions are increased in order to avoid citizens being involved only when they are ill or need health services (as in Germany, for instance) (OECD, 1992, 1995). On the whole the tendency is to curtail demand by acting on the supply, which is limited, cut or tied to financial co-participation.

This type of approach assumes the capacity to identify a precise system of relations among actors, to define the boundaries of this system and therefore, through adequate instruments, to 'direct behavior towards set objectives, on the basis of forecasts, given the existence of a governmental authority controlling the whole system of relations' (Bianchi, 1997, p. 276). The 'strong rationality' principle underlying this intervention is connected to the presence of a decision maker who has the power to set objectives, rank them and decide how to reach them and, especially, who is capable of a clear, consistent answer. This model seems to be typical of countries in which the state finances and provides health services, but also of systems of health insurance where the central authority is supposed to have the power to impose controls through less direct, more negotiated mechanisms.

At an organizational level all actions to be carried out to reach objectives are carefully planned. 'Making projects' uses words like information, efficiency, optimization, implementation, formalization, procedure, tasks, sequence and order; that is, concepts and instruments which are typical of a scientific organization of work based on a clear distinction between those who design and those who carry out policies (D'Angella and Orsenigo, 1999). In this situation the standardization of procedures should allow for the definition of programmes which, once defined, can be re-implemented without the need for further procedures and which can become stable cognitive and organizational routines.

However, a top-down policy like the one we are considering presupposes the existence of a 'closed entity, whose internal dynamics can be subjected to control so that a positive stimulus (incentive) or a negative one (prohibition), from a higher authority is necessarily followed by a corresponding move by the actors that make up the system under control' (Bianchi, 1997). In fact, the 1980s demonstrated not only that European countries and their economies are complex contexts, difficult to be reduced into units, but also that this complexity is particularly evident in the health context owing to the variety of actors involved and the nature of services provided. The presence of public and private actors, highly differentiated and interdependent, the continuous problems that service providers have to face in situations which are difficult to explain and manage, the high degree of involvement of all actors, all make the political and organizational contexts of policy implementation open to disorder, turbulence, unforeseen events and therefore unsuitable for hierarchical and highly rationalized forms of intervention.

As a matter of fact, setting aside the remarkable differences among nations, the analysis of policy implementation clearly shows the shifting of expenditure towards uncontrolled areas (for example, specialized medicine instead of hospital medicine) or towards the private market, with consequences such as less equity, users' growing dissatisfaction and a general process of downgrading of the public sector. Observers emphasize a gap between generalized control of costs and expenditure rationalization, so that very often a balanced budget does not coincide with a higher level of services (OECD, 1996).

Thus, in this early period, attention was focused on possible waste and the wrong use of productive factors (drugs, techniques, medical and hospital services and so on) as well as on the assessment of service efficacy and efficiency. 'Systematic evaluation' played a central role because priority was given to treatments whose efficacy seemed to be demonstrated and which could be widely applied. National and international bodies regulated the appearance of new drugs and new health techniques on the market more and more openly. Moreover, when innovations were made available to patients, their use was subjected to guidelines written by professionals.

Although prevention and health promotion remained basic values, the context was characterized by constant tension between administrators, who had to manage the health care system, and physicians, who were eager to experiment with new technical and scientific methods. For this reason budget constraints heavily affected therapeutical choices and the introduction of innovation was hindered or slowed down. It is not by chance that a large part of the debate stemming from these early policies, which remained almost unaltered in the following years, concerned essential prescriptions which should be guaranteed by public health systems and the degree and form of the state's responsibility. It is therefore possible to say, as some authors do, that the relationship between health policy and industrial policy was inexorably linked to the definition of the compatibility between economic resources available and attainable health levels (the HM model, according to Di Tommaso and Schweitzer, 2000).

4 A PUBLIC POLICY OF MANAGED COMPETITION

The second model of intervention on health expenditure tries to find balanced factors in a microeconomic plan. At first, interest is focused on the relationship between service providers (physicians, hospitals and so on) and actors responsible for health care (such as insurance companies and National Health systems), with particular regard to the hospital context. Later on two problems are faced: that of the physician–patient relationship in primary care and the problem of the reduction of pharmaceutical expenditure.

Great Britain, Holland, Sweden and Italy are the countries which decidedly follow this path. The first objective is the definition of the role of health funders who are given more responsibility in cost control and service evaluation. This means that, while on the one hand the health funder must find accounting forms of limitation, incentive and control, on the other he must institutionalize professional auditing methods by defining a whole series of best practices. In order to reach this goal, competition among providers is created by distinguishing the financing function from the production function.

This idea came about in the USA, where, in the 1980s, Alain Enthoven suggested abandoning the traditional health market model, based on patients' payments of fees for services and insurance reimbursement, in favour of a new relationship between the insured, insurers and service providers. In other words, insurers should directly enter the health process by guiding patients towards the facilities with which they have made an agreement. In this way, while the insured see their possibilities of choice greatly reduced, as they can only choose the insurer, the latter can limit costs by contracting with health centres a package of services which patients can get on payment of an annual amount. This new model is based on two levels of competition, one among purchasers of services, that is the insurance companies taking the name of health maintenance organization (HMO), and the other among providers, that is hospitals and specialized clinics (Maciocco, 1998, 2000).

This experiment was very successful in the USA and was introduced into Great Britain in 1991 with Margaret Thatcher's reform which, although leaving the basic principles of Beveridge's model unaltered, separated the purchasing and funding function of services from the productive function. The former, Maciocco says, 'was assigned to health authorities (the Italian ASL) which became purchasers, the latter to Trusts (hospitals and also agencies providing community services) which became providers. Trusts were granted complete managerial autonomy (before the reform, hospitals were under the administration of health authorities). The rationale for the reform was to create competition among providers inside the NHS, which explains the name of internal market or quasi-market given to the model. However, the 1991 reform laid the foundations of competition among purchasers as well since fundholding, a formula that allows even general practitioners to purchase services from trusts, gave rise to two potentially competitive actors on the contracting side' (Maciocco, 2000, p. 3). Moreover, the introduction of contractual relations and payment for services (DRG) into hospitals had the purpose of identifying costs, prices and services better. In the case of primary care this objective can be achieved by emphasizing the gatekeeper function of physicians and giving them a direct role in the choice of the services their patients need.[3] As for pharmaceutical expenditure, cost reduction should equally proceed from the involvement of those who prescribe drugs, as well as from the users' contributions to expenses.

Therefore, if the 'managed competition' model is designed and implemented in health systems of a universalistic type, where it is easier, given the presence of only one decision maker, to introduce large reforms, in countries with insurance health systems like Germany and Holland, competition among hospitals and insurance companies is fostered in the belief that it can better control expenditure while preserving the typical feature of this type of reform, that is the efficiency/efficacy equation.

In substance, the reform strategy is based on a criterion of 'non-choice' between a system entirely planned and controlled by the state and another dominated by the market. The model wants to combine the features of the two regulatory systems. More specifically, as attention is focused on the behaviour of different actors, an attempt is made at directing them through concepts and instruments, which condition their choices.

Taking the complexity of the system for granted, the aim is to implement 'limited rationality policies' where complexity is due to the environment while the problem-solving process is governed by the simple rules of competition and contracting out.

On an organizational plane, the problem is no longer to find the ideal structure and functioning, but the best solutions to problems as they present themselves. 'Problem solving' becomes, in other words, a way to decide on satisfactory solutions. In order to reach this goal it is necessary to understand the problem through the collection and processing of information, to resolve it into its components and to determine the logic of each of them as well as the specific knowledge necessary to face them. In this way it should be possible to plan action in different concrete situations, according to the real needs and the potentials of the actors involved, responding to environmental and organizational turbulence and uncertainty (D'Angella and Orsenigo, 1999).

The limitations of the above approach are to be found not only in the information and transaction costs involved in contractual relations between health funders and providers but also in the incorrect use of budgets, in the difficulty of reaching high levels of managerial and professional quality and, finally, in users' insufficient amount of control.[4] The introduction of sectorial budgets itself is not always related to the efficacy and equity of results.

More specifically, on an organizational level, the fragmentation of problems often overshadows the global objectives of the project. Actually it is not always possible to split a complex problem into smaller ones because the sum of the components is not equal to the whole. Moreover, as programming by problem solving involves several cognitive and emotional levels, it may happen that objective data are altered and transformed by subjective reactions. In other words this approach is based on one or a few decision makers, and underestimates interaction processes between various actors, which are common in organizational health contexts. From this perspective, planning must consider a complexity that is at the same time exogenous and endogenous and can hardly be faced by means of 'limited rationality' instruments (D'Angella and Orsenigo, 1999, p. 63).

In this second phase productive factors became the direct competence of service providers who could exploit, through them, their competition potential. Within the limits imposed by budgets, general practitioners and especially hospital administrators could increase their appeal, developing more services, improving the quality of the existing ones and introducing new diagnostic and therapeutical solutions. The increased managerial and organizational flexibility contributed to overcoming administrative bureaucracy typical of the public sector, which was highly centralized or dependent on decisions made at a national level. In this way the health industry in all its components became coherent to a model of health policy aimed at efficiency and customer satisfaction. The dilemma here is between the need for competition and the appeal to users, on the one hand and limited budgets, on the other, given the continuous introduction of new biomedical techniques and the restructuring strategies necessary for their diffusion. In other words this situation can produce both overproduction of services and duplication of individual treatment courses, together with a decreased interest in prevention and collective health promotion, as only health interventions with short-term social and economic benefits are privileged. Here Di Tommaso and Schweitzer's HIM model seems to be applicable as the potential of the health industry becomes more evident to funders and providers of health services who are, at least at a microeconomic level, in a position to exploit them, although within the limits of budgets adapted to users' and decision makers' needs. In other words the defensive attitude towards the different components of the health industry, typical of the late 1970s and early 1980s, seems to be abandoned, while the social and health capacity of development connected to it seems to gain more space.

5 PROGRAMMING BY INTEGRATING SETTING

In the late 1990s a third form of health policy appeared in the Scandinavian countries and Great Britain, and partly also in Italy. More specifically, in Norway and Denmark, decentralization of responsibilities made necessary coordination between institutional actors in charge of hospital care (entrusted to counties) and those responsible for primary care (entrusted to municipalities). In Great Britain, on the contrary, competition among purchasers (general practitioners and health authorities) required coordination between the budget for primary health care and that for hospital and community care (Light, 1997). In the latter case the Labour government suggested in 1997 the abolition of the internal market and the introduction of a system in which, although the distinction between purchasers and providers remained, cooperation took priority over competition among institutional actors.

As a consequence, 'managed cooperation' has the purpose of increasing coordination and integration among actors who, within a limited area, are involved in health service provision. The basic idea is that local authorities are the most suitable to manage health expenditure as they know both users' requirements and existing health facilities. Local authorities need to be supported in their search for interrelational forms ensuring people-continuity in health care and service decentralization. Emphasis is obviously on collaboration rather than on competition, in a subsidiarity logic of responsibility and medium-level control (OECD, 1996). As a matter of fact, while competition tends to give rise to unplanned medium-term contracts and to erode the values underlying health assistance and care, cooperation has a longer-term perspective and takes place in a general context of planned actions and requires common objectives and values (Borzel, 1998). This does not exclude contractual and financial relations for profit seeking by each partner (Light, 1997); however, it is specialization that triggers *complementary searching* through the identification of social representations and common systems of value.

On an organizational level it becomes essential to build a common vision of problems and actively involve all actors in the development of possible solutions. Equally important is the evaluation of results and the research-action activity in a process which is continually questioned and redefined. Integration and connections imply a relational communicative competence which is much more important than individual competence: it is the former which allows the development of common meanings and jointly planned actions, as well as adjustments in which dynamic processes are guaranteed and locked-in situations of conflict without growth are avoided.

With this model of 'dialogic planning' the linear system or the circular one, typical of previous health policies, would be abandoned. It could be possible, instead, to enter networks made of small or big entities, result-oriented, self-regulated and able to co-operate with each other. Obviously the autonomy of each entity does not exclude imbalances in power, influences and advantages; it only enhances the energy of each component and its individual capacity to cope with problems (Butera, 1990). However, models in which links are weak bring about a lack of roles, as well as the absence of a strong identity or procedures to follow. For this reason changing dynamics, the continuous confrontation of entities and the viability of each actor transform each encounter into a microcosm where rules must be learned and outcomes must be forecast.

To achieve this purpose, a plurality of vision would help identify the problem and find its solution through the development of forms of planning consistent with an uncertain, ambiguous and not always dynamic context (D'Angella and Olivetti Manoukian, 1999). Planning ceases to be an external 'a priori' situation, or a marginal condition as in the two previous models, and becomes an active situation in which actors can only reduce uncertainty and learn by doing through direct participation. There is therefore no dichotomy between thought and action, between order and control; instead the intervention implies at the same time knowledge and research. A trust relationship between the actors involved becomes essential. This trust is built in the field and represents the foundation of any co-ordination and integration process (Olivetti Manoukian, 1999).

As a matter of fact, the choice to carry out the control of expenditure and quality of services at a local authority level and the decision to make them depend on agreements and transactions which are not only economic and authoritarian but also ethical and social seem to imply the existence of actors and social networks on which to graft both institutional and particular trust relationships (Roniger, 1988, 1992). This situation should increase cohesion and motivation and limit opportunistic behaviour, besides increasing the control activity of the network itself and the socio-political community of belonging (Borzel, 1998). In plain language, the closeness of relations (personal, empathetic, rooted in local contexts and so on) should become the source of social control, complementary, not opposed to, institutional and public control (de Leonardis, 1999).

In the case of these integration-oriented public policies it can be assumed that the attention to the potential of the health industry remains constant, especially where the specialization of actors in the production of services remains unaltered. However the need for coordinated and integrated projects, based on common values and strategies, could bring about a tighter control of overproduction risks and a lack of interest in collective health typical of previous policies. Above all, if these forms of managed cooperation remain inside a community welfare framework, directed by public interests, it will be possible to keep excessive consumerism and therapeutical individualism under control through forms of network regulation. More specifically the health industry might be asked to demonstrate the same level of collective ethics and public service spirit required of health professions in their working activities. This means, perhaps, that in order to be consistent with these last forms of health policy the Health Industry Model will have to show a greater capacity for integration into the system, not only in economic terms but also in ethical and social terms (Health Industry Integration Model: HIIM).

6 TOWARDS NEW REGULATORY FORMS

It would be wrong to think that the changes which have taken place in European health policies demonstrate evolution or convergence towards similar forms of regulation. It would be too simplistic to imagine a shift first from contributive to redistributive modalities and then a further step towards community provision. Even if this tendency exists, it is accompanied by the experimentation of forms of regulation stemming from the encroachment and coexistence of models once deemed to be antithetical.

As is well known, welfare regulation by the state and the market represents the basis of a model which, starting from the one described by Titmuss, was used to differentiate national welfare and health systems (Esping-Andersen, 1990, 1999). On this basis these have been identified as market-dominated systems like the American one, state-dominated systems like the British one and also regulated mixed systems, as in Germany and France, by actors representing the interests of society, although within a regulatory framework guaranteed by the state (compulsory insurance) (Martinelli, 1983).

This first classification was followed, at least in the Italian debate, by the need to introduce a *tertium genus*, a third kind, not a hybrid but a substantial one. Referring to a previous study by Trigilia and especially to Polanyi's model of transactive modalities characterizing the distribution of resources in society, Massimo Paci proposed, in the late 1980s, the introduction of reciprocity as a third way: 'In an exchange appropriative movements are bilateral and take place within a market system; in redistribution they move from the social periphery to the institutional center and vice versa according to the level of authority. In reciprocity, instead, the allocation of resources is symmetrical and takes place within a solidarity system or a system of tight bonds inside the community' (Paci, 1989, p. 34). This means introducing into the analysis of welfare systems not only the state and the market but also community institutions in their traditional or most advanced forms. He refers to the protection provided by the family, parental networks and the neighbourhood to self-help groups or voluntary organizations. In this case integration does not stem from exchange or hierarchy, as in the previous models, but from the sharing of values and orientations, which imply a large consensus on present and future actions.

The reference to Polanyi appears to be obvious not only for the models of socioeconomic integration but also for the idea that different forms of integration can coexist inside the same society and also if one of them dominates. In relation to this, Paci suggests an analysis of national systems of welfare focusing on a mix of allocative–regulatory functions that dominates in a different context and can be defined as a dominance mix. The geometric image of the triangle makes it possible to describe models of welfare in which the main role is sometimes played by exchange, sometimes by redistribution and sometimes by reciprocity (Figure 3.1).

This static image of welfare systems is transformed by Paci (first) and Ferrera (later) into a dynamic interpretation aimed at defining the forms of change. The hypothesis according to which there exist historical models of

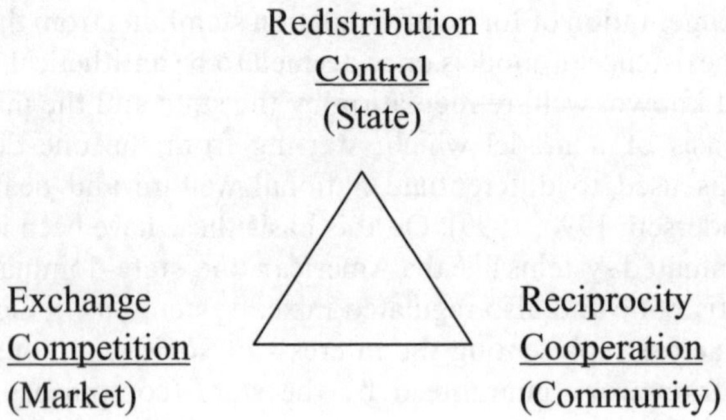

Figure 3.1 Graphic representation of the three traditional forms of social regulation

dominance welfare assumes that one model can replace the other and that, in the history of a single country, different models alternate and reappear in the course of time. The theme of change leads Paci to assume the existence of long waves in the trends concerning the systems of welfare and therefore historical recurrences. Ferrera, instead, is in favour of a model which takes into account the changes which have occurred within a national context, distinguishing purely universalistic and occupational systems of welfare from mixed ones. The former model applies to France, Germany and the Scandinavian countries presenting historical continuity in their regulations; the latter pertains to Italy, Holland and Great Britain, that is to say to countries where models of social protection have changed over the course of time (Ferrera, 1993).

On the basis of these indications it is possible to conclude that, in the 1980s and 1990s, many European countries passed from a redistribution dominance welfare mix to an exchange dominance one, while at the moment 'reciprocity' appears to be the key word. However, a closer analysis of health policies shows more complex forms of mix in which regulations stem from the integration of traditional systems. *Managed competition* would be nothing but a mix of private and public allocation, as the idea is to make public agencies and facilities face competitively the regulatory mechanisms of the market. On the other hand, *managed cooperation* seems to be the integration of redistributive and reciprocal forms of regulation, because it presupposes the existence of public facilities at a local level while insisting on the need that they collaborate with each other and with private organizations sharing objectives and strategies. A third way is that of *competition–cooperation*, an expression which seeks to underline the presence of instruments which, although connected to market logic, have a no profit aim but only ethical or humanitarian princi-

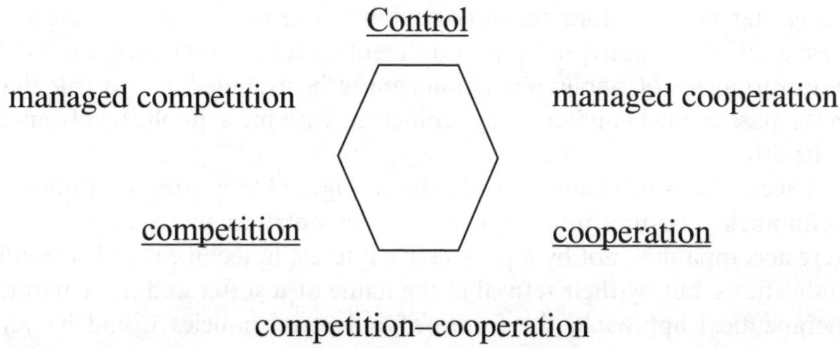

Figure 3.2 *Graphic representation of new and traditional forms of social regulation*

ples. This area, often identified as the third sector, includes both traditional and innovative forms because it comprises both social insurance arising from the transformation of self-help organizations and more modern types of non-profit organizations.

In conclusion, dominance systems of regulation with at least six allocation mechanisms could be identified, so that a hexagon (Figure 3.2), rather than a triangle, should be used to represent them. As a matter of fact, if competition, cooperation and control represent the traditional poles of integration (corresponding to the market, the community and the state), managed competition, managed cooperation and competition–cooperation could find a place in the middle of a continuum between the two traditional forms.

7 CONCLUSIONS

The paths we have outlined in relation to the existing link between health policies and sectorial industrial policies could change in the short term because new regulations might emerge, mainly centred on the market. Some authors, for instance, see the evolution of health systems as an axis of socioeconomic change, which would involve on the one hand 'community rationalization' and on the other 'the liberalization of the health market' (Paccaud, 2001). If this was the path taken by European countries, economic profitability would take the place of social and political profitability completely in health service provision, and this would obviously affect the equity of treatments and the role of professionals, patients and even health facilities. In view of future scenarios of increasing deregulation, technological innovation choices are likely to have a different nature and take a different direction. However, a possible return to forms of public regulation cannot be excluded, if it is true that, in the past, armed conflict always coincided with more public involvement in health.

A second axis of change can also be configured, opposing scientific and technological optimism to scepticism. If deregulation and recentralization were accompanied, not by a progressive interest in technical and scientific innovations, but by their refusal in the name of a softer and more natural therapeutical approach, the future of industrial policies would be very different.

In conclusion, leaving future scenarios aside, focusing on health and industrial policies in their temporal and spatial contexts seems to be necessary. This awareness must support any kind of analysis, both retrospective and exploratory, able to highlight the complexity of history and of different contexts, together with the simplicity of ideal-type models, both to be read in their dynamic and constructive relationship.

NOTES

1. The data concern OECD countries in Europe.
2. In nominal values, between 1966 and 1995 health expenditure in Italy increased by 153 times, while in real values it grew by 4.6 times, with an average rate increase of 4.5 per cent per year. This means that the quantity of services and goods consumed grew more than the GDP, which in the same years increased by 3.3 per cent per year (Mapelli, 1999).
3. According to LeGrand the introduction of managed competition into Great Britain makes this system at the same time similar to and different from the American one. General practitioners become somewhat similar to the American HMO because they can contract a package of services for their patients with hospitals and specialists. Hospitals become similar to the corresponding American non-profit agencies as they are organizationally independent but not oriented towards earnings and profits to be distributed. However, together with these similarities there are differences due to the funding being in the hands of the state which contributes to over 90 per cent of the expenditure. Patients can choose their general practitioners but not the health authority to which they belong according to residence. Hospitals, although independent, remain public and both purchasers and providers see their freedom limited and controlled by statutory administration (Le Grand, 1999).
4. In the case of Great Britain, LeGrand points out the difficulty in evaluating managed competition, as no assessment was made by the Conservative government. However, the existing indicators seem to point to increased efficiency and greater responsibility of providers due to the new role of general practitioners. These data should, however, be interpreted in the light of increased administration costs, the risks faced by patients in selection and, finally, users' diminished freedom of choice. The difficulty of properly evaluating data depends, according to the author, on the fact that the NHS has maintained its centralized and public role in spite of the 1991 reform. Besides, the actors do not seem to have identified themselves in the competition culture as they continue to be service-oriented. Moreover the process of contracting was partially hampered by difficulties due

to the specialization of services requiring mutual trust, professional discretion and long-term cooperation (LeGrand, 1999).

REFERENCES

Bianchi, P., 'Nuovi approcci alla formulazione di politiche pubbliche in situazioni di complessità, integrazione europea, politica industriale, risorse umane', *Economia & Lavoro*, **3–4**, 1997, 273–80.

Borzel, T.A., 'Le reti di attori pubblici e privati nella regolazione europea', *Stato e Mercato*, **54**, 1998, 389–432.

Butera, F., *Il castello e la rete. Impresa, organizzazione e professioni nell'Europa degli anni Novanta*, Milan: F.Angeli, 1990.

D'Angella, F. and Olivetti Manoukian, F., 'Ascolto e osservazione nella progettualità dialogica', *Quaderni di animazione e formazione*, 1999, 83–97.

D'Angella, F. and Orsenigo, A., 'Tre approcci alla progettazione', *Quaderni di animazione e formazione*, 1999, 53–68.

de Leonardis, O., 'Terzo settore: la doppia embeddedness dell'azione economica', *Sociologia del Lavoro*, **73**, 1999, 230–52.

Di Tommaso, M.R. and Schweitzer, S.O., 'The health industry: more than just containing costs', *L'Industria – Rivista di Economia e Politica Industriale*, **3**, 2000, 403–26.

Esping-Andersen, G., *The Three Worlds of Welfare Capitalism*, Princeton: Princeton University Press, 1990.

——, *Social Foundations of Postindustrial Economics*, Oxford: Oxford University Press, 1999.

Ferrera, M., *Modelli di solidarietà. Politica e riforme sociali nelle democrazie*, Bologna: Il Mulino, 1993.

Le Grand, J., 'Competition, cooperation, or control? Tales from the British National Health Service', *Health Affairs*, **XVIII**(3), 1999, 27–39.

Light, D.W., 'From managed competition to managed cooperation: Theory and lessons from the British Experience', *The Milbank Quarterly*, **LXXV**(3), 1997, 297–341.

Maciocco, G., 'La terza via di Tony Blair', *Prospettive Sociali e Sanitarie*, **7**, 1998, 1–5.

——, 'Le riforme dei sistemi sanitari nei paesi industrializzati', mimeo, 2000.

Mapelli, V., *Il sistema sanitario italiano*, Bologna: Il Mulino, 1999.

Martinelli, A., 'Salute e sistemi sanitari occidentali', in P. Donati (ed.), *La sociologia sanitaria*, Milan: Angeli, 1983.

Mossialos, E. and LeGrand, J., *Health Care and Cost Containment in the European Union*, Aldershot: Ashgate, 1999.

OECD, 'The reform of health care. A comparative analysis of seven OECD countries', *Health Policy Studies*, **2**, 1992.

——, 'Next directions in health care policy', *Health Policy Studies*, **7**, 1995.

——, 'Health care reform. The will of change', *Health Policy Studies*, **8**, 1996.

——, 'Health outcomes in OECD countries: a framework of health indicators for outcome-oriented policy making', Labour Market and Social Policy Occasional Paper no. 36, Geneva, 1999.

Olivetti Manoukian, F., 'Generare progettualità sociale', *Quaderni di animazione e formazione*, 1999, 5–11.

Paccaud, F., 'Gli scenari futuri dei servizi sanitari', *Salute e Territorio*, **126**, 2001, 87–100.
Paci, M., *Pubblico e privato nei moderni sistemi di Welfare*, Naples: Liguori, 1989.
Pierson, P., 'Lo stato sociale nell'era dell'austerità permanente', *Rivista Italiana di Scienza Politica*, **XXIX**(3), 1999, 393–440.
Roniger L., 'La fiducia. Un concetto fragile, una non meno fragile realtà', *Rassegna Italiana di Sociologia*, **3**, 1988, 383–402.
——, *La fiducia nelle società moderne. Un approccio comparativo*, Messina: Rubbettino, 1992.

4. A hedonic model of pricing of innovative pharmaceuticals*

William S. Comanor, Stuart O. Schweitzer and Tanja Carter

1 INTRODUCTION

The Health Industry Model, as described in Chapter 1 of this volume, suggests that investment in health is useful, not only for its beneficial effect on a population's health status, but also because of its positive effect on health-related research and development and spillovers that benefit other high-technology sectors. Investment in research and development, however, depends upon market conditions affecting the potential fruits of these investments. Investment in R&D will be undertaken only if the investor expects that a new product, once brought to market, can be sold at a price that will cover its costs of development as well as its manufacturing costs. Without this expectation, there is little incentive to develop innovative products. In the case of pharmaceuticals, the market is essentially worldwide, even though it is dominated by the industrialized countries. In principle, prices in a particular country should matter to a seller only in proportion to that country's worldwide market share. Nevertheless there appears to be a correlation between pharmaceutical prices in a country and the level of its pharmaceutical research and development activities.

This discussion points out the importance of prices in influencing the investment in innovation in the pharmaceutical industry, both in the aggregate and in particular countries. In many countries, drug prices are set by governmental or quasi-governmental agencies, often with reference to *other* prices, either of comparable products in the particular country or of the same product in other countries. But even in countries where prices are set more by market forces, the process of price determination is not well understood. The industry often claims that prices are determined by the costs of R&D: a supply-driven model. Although the full costs of production (including costs of R&D) must be covered in the long run for a firm to remain viable, this explanation is insufficient to determine the prices of particular products. What other factors determine pharmaceutical prices in a market-driven setting?

This chapter explores the relative importance of both market factors and therapeutic characteristics for the prices of drugs sold in the United States. Since the incentive to undertake costly research and development on new pharmaceuticals depends largely on the prices for which the resulting products can be sold, this study is an important component in a larger effort to determine the factors which give rise to these incentives. What is relevant about the decision to invest in pharmaceutical research is not merely the cost and risk of developing new products, but also the prospective returns that follow once new products have been developed. This analysis contributes to our understanding of this second set of factors.

The prices paid for pharmaceuticals in the USA depend not only on the therapeutic characteristics of the products but also on the nature of the buyers. Indeed there is considerable evidence that certain large buyers pay much lower prices than do individual customers who must pay cash prices. This study explores these differences and relates them to the therapeutic properties of the drugs being sold. It provides empirical results for five specific classes of pharmaceuticals.

2 PRESCRIPTION DRUG PRICING

The Cost of Drug Development

Bringing new drugs to market entails high initial fixed costs for research and development, and also for promotion (Comanor, 1986). Kettler (1999) estimates the total cost, in 1997, of bringing a new drug to market at \$312 million.[1] Approximately half of this sum represents actual research and development outlays, with the rest being opportunity costs resulting from the considerable delay between outlays made and revenues received.

After a drug is approved for commercial release, significant marketing costs are generally required. Their purpose is to inform physicians and other health care professionals of the existence and therapeutic potential of the new product. Traditionally, pharmaceutical manufacturers have relied on a sales force to promote drugs to individual physicians as well as on advertising in medical journals. With the growth of managed care, marketing is also directed towards the pharmacy committees of those organizations. In addition, the FDA (Food and Drug Administration) permits direct-to-consumer advertising through television and other media. Marketing outlays are frequently substantial immediately after a drug's release but often decline over time, along with the product's sales within its therapeutic class. Production costs vary across drugs, but generally these costs comprise only a minor proportion of product revenues.[2]

Market Demand

Largely because of the patent protection on new products, pharmaceutical manufacturers of branded products are insulated to some degree from competition.[3] However, the size of a drug's mark-up over production costs depends on the elasticity of demand by those who purchase the products. When only a few comparable substitutes exist, demand is relatively inelastic, and a high price can be commanded in the market without significantly reducing sales. On the other hand, when substitute therapies are abundant, demand is generally more elastic and purchasing decisions are more responsive to changes in price. The availability of substitutes limits the ability of drug makers to increase price without experiencing a marked decline in sales (Lu and Comanor, 1998).

In the USA, nearly half of all prescriptions are filled by generic products (Teitelbaum, 2003, p. 6). In contrast to branded products, generics are largely sold by smaller manufacturers who undertake little research or few marketing efforts. There is little differentiation among these products, and prices typically approach marginal costs when there are a sufficient number of firms supplying the product.[4]

Purchasing Groups

Pharmaceutical firms typically discriminate across buyer groups since demand elasticities can vary significantly. As a result, there is generally no set price for a product independent of the particular buyer. Firms maximize profits by exploiting differences in the elasticities of demand across groups of buyers.[5] In particular, there is a critical difference between health maintenance organizations (HMOs) and pharmacies which arises from an HMO's ability to determine which drugs are used and which are not in contrast to a pharmacy's role of largely filling prescriptions that are written by others.[6]

This distinction has become particularly important in recent years. By 1996, nearly 59 per cent of retail pharmacies' revenues from drug sales were at least partly paid by managed care coverage. Cash payments represented 29 per cent, with the rest largely accounted for by Medicaid payments (CBO, 1998, pp. 7–8).

In the analysis below, we assume that HMOs and pharmacies are the relevant buyers in the wholesale marketplace, and that the pharmaceutical companies are the relevant suppliers. Although most products are physically supplied by retail pharmacies, these firms act as agents for the HMOs for that segment of demand.

3 HEDONIC THEORY

According to hedonic theory, consumer valuations depend on a drug's positive and negative therapeutic attributes.[7] Drugs indicated for similar conditions, which 'bundle' together different benefits and risks, should therefore differ in price. For example, drugs that are more effective for treating particular conditions should command higher prices from consumers than those demonstrated to be less effective, *ceteris paribus*. To the extent that a drug is associated with more serious side-effects than an alternative, patients/physicians should be willing to pay less for it than for the other. Similarly products that are identical in all relevant aspects should sell for the same prices.

The value of drug characteristics is not directly observable because individual characteristics are not traded directly in the marketplace. However, the implicit values of therapeutic attributes, or 'hedonic prices', can be estimated by observing the variation in prescription prices and drug characteristics for a given therapeutic category. What is then determined is the marginal valuations of the individual characteristics.[8]

If the price of a prescription drug is related to the magnitude of its various characteristics, we can express this relationship as the following:

$$p_j = f(z_{1j}, z_{2j}, \ldots, z_{mj}), \tag{4.1}$$

where p_j equals the natural logarithm of the price of product j and z_{ij} is the value of the ith characteristic of the jth drug, where $i = 1\ldots m$. The market price should then reflect the hedonic value of the drug's attributes. We consider three classes of attributes: therapeutic benefits, side-effects and convenience. Favourable attributes are expected to add positively to a drug's market price, and negative attributes to detract from it. As is conventional, we assume a linear form for this expression:

$$p_j = \beta_0 + \beta_1 z_{1j} + \beta_2 z_{2j} + \cdots + \beta_m z_{mj} + e. \tag{4.2}$$

The price effects of drug attributes are then derived from the first derivative of the hedonic function so long as the z_{ij} are continuous variables:

$$\partial f / \partial z_{jj} = \beta_i \qquad i = 1\ldots m, \tag{4.3}$$

where β_i represents the additional outlays required to obtain a product variant with one more unit of the particular attribute. Where the z_{ij} are discreet variables, the β_i describe the additional spending required to obtain this particular characteristic.

The conventional assumption required for hedonic coefficients to represent the prices of the relevant characteristics is that product prices are determined competitively. Although pharmaceutical prices for branded products typically exceed marginal costs by substantial margins, these differences are much lower, if present at all, for generic drugs which are generally supplied competitively. These products serve as the base case from which other prices are compared. The market factors included in the equations are therefore essential elements in the empirical analysis.[9]

4 HYPOTHESES TO BE TESTED

This study measures the contribution of both product attributes and market factors to the prices paid for pharmaceuticals by HMOs and pharmacies. We introduce hedonic analysis to determine the role played by the three product characteristics: therapeutic benefits, the severity of any side-effects and the convenience to patients of using the product. Among the market factors that we investigate are brand-name effects, type of purchasing organization, the availability of generic alternatives and the relative novelty of the product as determined by its 'generation'. Following hedonic theory, drugs with greater efficacy should command higher prices. Furthermore those with more serious side-effects should be associated with lower prices; also patients should be willing to pay more for drugs offering more convenient dosage regimens.

According to FDA regulations, branded and generic drugs must have the same active ingredient and be 'bioequivalent'. An even more rigorous standard is 'therapeutic' equivalence, indicating equal therapeutic effects of the drug. Even when generic products are *therapeutically* equivalent to their branded counterparts, branded drugs are often viewed as superior by consumers. In part, this is because brands create an 'image' associated with quality and/or innovation. Patients may perceive a quality difference even where no substantive difference exists. They may also be 'brand loyal' to original products long after generic substitutes become available. When insured patients are asked to pay the same nominal price for branded products as they do for generics, branded products are generally preferred. Even when insurers require greater payment for brands than for generics, some patients will refuse to switch to the generic product with which they have no personal experience. We thereby expect higher prices for brand products than for generics, *ceteris paribus*.

For the reasons suggested above, we anticipate that HMOs will pay less than pharmacies for branded products. However, there has been little consideration given to whether HMOs will also pay less for generic drugs, which is also examined in the analysis below. We also test whether the different purchasers respond in the same way to the various product characteristics.

Hedonic theory predicts that product attributes determine price. However, measured differences in attributes among competing products are often relatively small. We therefore introduce another variable: the 'generation' of the product. We hypothesize that later generations of drugs are frequently *perceived* as improved in ways that may not be captured by the measured attribute variables. More recent product generations may therefore lead to higher prices, except where physician prescribing patterns are particularly difficult to change.

5 AN ESTIMATION ISSUE

As Epple (1987) points out, hedonic equations are typically subject to endogeneity problems since consumers generally select prices and product attributes at the same time. This simultaneity is reflected in the estimated demand functions; he writes, 'it is present whether the distribution of characteristics supplied in the market is exogenous or endogenous' (ibid., p. 64). The result of this problem is that estimates derived from ordinary least squares (OLS) are inconsistent.

While this critique certainly applies to final consumers who must choose from among the various products available, it does not apply to wholesale buyers who instead purchase all available products at the same time. The drug manufacturers determine the characteristics of their products subject to the constraints imposed by technology and FDA approval requirements. Prices are subsequently determined on the basis of these characteristics and market conditions (Lu and Comanor, 1998). The wholesale buyers then react to both the prices set and product characteristics in making their purchasing decisions. As a result, we effectively have a recursive model in which the buyers' willingness to pay for particular attributes affects the prices that can be charged, but there is no reverse causality. And, in that case, the OLS estimates are consistent.

6 THERAPEUTIC CATEGORIES

Meaningful assessments of therapeutic effectiveness can only be made by comparing drugs within the same therapeutic category. This study focuses on drugs in five therapeutic categories: antidepressants, calcium channel blockers, antihistamines, antimicrobials and glaucoma drugs.

Antidepressants

Antidepressants are used primarily to treat depression and anxiety in psychiatric patients. Drugs in this category include heterocyclic antidepressants (HCAs), monoamine oxidase inhibitors (MAOIs) and serotonin/norepinephrine reuptake inhibitors (SSRIs).

Calcium Channel Blockers

Calcium Channel Blockers (CCBs) are used to treat hypertension, angina and other cardiovascular disorders. Their action is designed to stop the movement of calcium ions into cardiac and vascular muscle. CCBs help dilate coronary arteries and peripheral arteries and can also slow the heart rate. Drugs within this category vary therapeutically according to the manner by which they have an impact on vascular smooth muscle alone or the vascular muscle and myocardium together. Common side-effects include dizziness, headaches and nausea.

Antihistamines

Histamine H1-receptor antagonists suppress allergic responses to histamine, including seasonal hay fever, allergic rhinitis and conjunctivitis. Histamine is an arteriol dilator that, when released, can cause various effects, including increased salivary and bronchial gland secretions. Antihistamines block the effect of histamine, which increases the permeability of capillaries and stimulates sensory nerves. These drugs generally relieve symptoms such as sneezing, watery eyes and runny nose, and may minimize breathing difficulties. Common side-effects include drowsiness and dizziness.

Antimicrobials

Antimicrobial agents kill or prevent the reproduction of organisms, preferably with little or no toxic effect on the host body. They are derived from either bacteria or moulds, or are produced chemically. Combinations of antimicrobials are often necessary to treat serious infections. While broad spectrum antibiotics kill or impede a wide range of Gram-positive or

Gram-negative bacteria, narrow spectrum antibiotics are largely effective against either Gram-positive or Gram-negative bacteria, and so are more effective against a single organism or disease.

Ophthalmic Solutions for Glaucoma

Glaucoma is a condition of increased fluid pressure inside the eye (intraocular pressure) and occurs when the fluid within the eye (aqueous humour) drains improperly. This process damages the retina and optic nerve, causing partial vision loss, and can eventually lead to blindness. Glaucoma is a leading cause of blindness. While there is no prevention available for the development of glaucoma, further vision loss and blindness may be averted with appropriate treatment.

7 DATA

Prices

The pharmaceutical prices paid by HMOs used in this study are those actually paid by a leading HMO in California during the fourth quarter of 1997, as developed from company records. They include any rebates and discounts that were available. These prices are determined in the following manner: the total amount paid for each drug between 1 October 1997 and 31 December 1997 is divided by the number of units purchased for its most frequently prescribed strength. These prices are then converted to daily dosage prices at the usual adult dosage as specified in *The Medical Letter*, vols 38 and 39 (Abramowicz, 1996, 1997). For example, the price of a 100mg tablet was multiplied by three to arrive at the price per day of a drug with a usual adult dosage of 300mg per day. Data on prices paid by the HMO for generic drugs are non-specific as to manufacturer.

Pharmacy prices are the wholesale prices paid by pharmacies to a major drug wholesaler in December 1997, and thereby reflect the prices paid by cash customers. They do not include any dispensing fees or any discounts, typically small, that individual pharmacies may have received. As with the HMO prices, pharmacy prices are converted to the price per usual adult dosage. In this case, generic prices for individual manufacturers are available.

Drug Attributes

Therapeutic effectiveness, side-effect profiles and convenience ratings are determined on a five-point scale with the assistance of physicians expert in the particular clinical area. Drugs are rated on effectiveness and convenience, with higher scores indicating improvements. The severity of side-effects is also ranked but here higher scores indicate more serious side-effects.

Five-point scales are effectively cardinal measures. They impose a condition of proportionality which suggests that drugs rated on efficacy, say, with a '5', are five times as effective as drugs scored with a '1'. In alternative regression specifications, these scales are replaced with the five-point ratings compressed into two-point, dichotomous, scales. For those equations, '0' represents ineffective products, and also those with mild or minimal side-effects, while a convenience score of '0' refers to relatively inconvenient dosing.

For antidepressants, therapeutic effectiveness is measured by the drug's ability to alleviate geriatric agitation, which is one of the most common uses of these drugs. Although antidepressants are of course used for other indications as well, we assume here that the drug's price is dependent on its success in treating this common condition. Other therapeutic variables in this class are highly correlated with this factor, so it can be considered a proxy for the general therapeutic benefits of antidepressants.

For antimicrobials, three therapeutic benefit variables are considered: the STREPB variable measures a product's ability to fight streptococcus B, the URI variable measures the ability to fight upper respiratory infections and the UTI variable measures effectiveness against urinary tract infections.

The therapeutic effectiveness of antihistamines is measured by the drugs' effectiveness against allergy symptoms. For calcium channel blockers and glaucoma drugs, the products are largely equivalent in terms of therapeutic benefits. So this variable could not be estimated, leaving side-effects and convenience as the relevant independent variables.

The severity of side-effects is again rated on a five-point scale, where higher scores indicate drugs associated with more serious side-effects. Convenience to patients was also rated on a five-point scale, with a higher score indicating easier compliance. For the most part, convenience is determined by the typical daily dosing regimen.

Market Factors

Market factors reflect the different conditions of sale, and are all represented by dichotomous (0–1) variables. The first variable is the distinction between branded and generic products. Branded drugs are designated by '1' and generics by '0'. The next two variables distinguish between the prices paid by the two classes of buyers, separately for branded and generic products. For the first of these variables, HMO purchases of branded products are indicated by '1', with '0' for all other observations. HMO purchases of generics are indicated by '1' with '0' otherwise. In effect, these three dummy variables describe the relative prices for the four underlying categories: branded products purchased by both HMOs and pharmacies, and generics purchased by both HMOs and pharmacies.

The final variable of this set represents the effects of generic competition. Branded drugs facing competition from generic products have a value of '1'. Branded drugs with no generic competition and all generic drugs have a value of '0'.

Generation

The generation dummy variables represent generations of drugs, which typically reflect important differences in therapeutic actions or results. To a large extent, a drug generation is a substitute measure for drug efficacy. In the case of antidepressant drugs, generation designations are from Lacy (1998–99). First-generation drugs are tricyclic antidepressants and monoamine oxidase inhibitors (MAOIs). Second-generation drugs, such as bupropion and trazodone, are largely cyclic antidepressants, while third-generation drugs are selective serotonin reuptake inhibitors (SSRIs).

Drugs in the other therapeutic categories were designated as first, second or third-generation drugs, depending on their date of introduction. All drugs introduced prior to 1982 are assigned to the first generation, drugs introduced between 1982 and 1990 are grouped into the second generation, and drugs introduced after 1991 are classified as third-generation drugs. To simplify the analysis, the same periods are used for all classifications.

8 ECONOMETRIC RESULTS BY THERAPEUTIC CATEGORY

We estimate hedonic regression equations for each drug category by ordinary least squares, using the traditional semi-logarithmic functional form (Griliches, 1961, 1971). This specification implies an increasing market price per unit of desirable characteristic. Furthermore the coefficients of the attribute variables describe the approximate percentage change in the relevant prices associated with a shift from one classification to another.

Tables 4.1–4.5 present the quantitative results for each therapeutic category. T-statistics are shown below each coefficient estimate. Coefficients marked (*) were significant at the 5 per cent level (one-tail test) and those marked (**) were significant at the 1 per cent level (one-tail test).

Antidepressants

Table 4.1 presents several regression specifications for antidepressant medications. Data on 14 distinct chemical entities are used to estimate these equations. Equation 1 shows that efficacy and side-effect profiles are both highly significant, with their effects on price in the expected directions. The coefficient of the third attribute variable, convenience, is insignificant and therefore dropped from the equations. An improvement in efficacy by one unit increases price by $1.57 per day on average, and a worsening of side-effects by one unit lowers the average price of these drugs by $1.33 per day. Branded products sell for significantly more than generics, although HMOs pay relatively more for generics than do pharmacies.

One possible explanation for this result can be found by considering HMO demand for brands versus generic alternatives. Managed care organizations may well have more elastic demands for branded products than do pharmacies because of their closely monitored formularies, which mandate switching to therapeutically equivalent generic alternatives wherever feasible. However, the demand for generics by HMOs may be more inelastic than the demand for generics by pharmacies because of their strong profit motive to reduce prescription costs by switching away from branded products.

Second-generation drugs are not more expensive than first-generation products, while third-generation products are *less* expensive, by an average quality-adjusted daily cost of $1.67. The overall R^2 for this specification is 0.95. As indicated in the notes to Table 4.1, equations 4 and 5 are re-estimated using dichotomous versions of the attribute variables. The results are quite similar to what was presented before, except that the second-generation effect is now positive and statistically significant.

Using the Wald test (Kmenta, 1986, pp. 492–4), we examined whether the coefficients for the variables included in equation (1) are the same as between HMOs and pharmacies. For all variables, except the branded versus generic distinction, we could not reject the null hypothesis that the relevant coefficients are equal. In addition, we examined the residuals from equation 1 in this table. Using White's test for heteroscedasticity (Kmenta, 1986, pp. 295–6), we rejected the null hypothesis that the variance of the error terms is constant throughout the distribution. However, when this equation was re-estimated to

Table 4.1 Antidepressants: factors determining 1997 prices paid by HMOs and pharmacies (dependent variable: natural log of price paid for a usual daily dose)

	Intercept	Geriatric agitation	Side-effect profile	Convenience	Branded product
1.	−2.084**	0.451**	−0.285**	—	2.794**
	(9.37)	(9.93)	(3.41)		(21.8)
2.	−1.806**	0.402**	−0.367**	—	2.418**
	(7.29)	(9.94)	(4.07)		(16.5)
3.	−1.918**	0.442**	−0.294**	0.065	2.813**
	(4.46)	(8.99)	(3.39)	(0.45)	(20.7)
4.	−2.200**	0.909**	−0.258**	—	2.807**
	(21.4)	(9.02)	(2.45)		(21.3)
5.	−2.165**	0.956**	−0.165	—	2.729**
	(17.9)	(8.22)	(1.47)		(16.8)

Notes on antidepressant equations:
1. Therapeutic properties, side-effect profiles and convenience aspects were determined with the assistance of Ralph Bien, MD.
2. All values were provided on a four-point scale. In equations 4 and 5, cells were combined to create a more conventional two-point scale.
3. Price data were obtained from two sources: (a) pharmacy prices were derived from the price lists distributed by a major drug wholesaler and applicable to the fourth quarter of

correct for this problem, there were only minor changes indicated for the *t* statistics and no substantive differences in the empirical findings.

Calcium Channel Blockers

The results for calcium channel blockers are presented in Table 4.2. All of the products in this class were generally equivalent in terms of therapeutic benefits, so no coefficient for the therapeutic effectiveness variable could be estimated. Instead drugs are differentiated on the basis of side-effects and convenience. The coefficient for the side-effect variable is unexpectedly negative and generally significant. From the coefficients estimated in equation 2, we find that more pronounced side-effects, when increased by one unit, lead to higher average prices by \$1.74 per day. The convenience coefficient is positive, as expected, and significant, indicating that improving convenience by one unit increases average price by \$1.64 per day.

While the coefficient of the HMO brand variable is not significant for this class of drugs, the results again suggest that HMOs tend to pay more for

HMO brand	HMO generic	Second generation	Third generation	Generic competition	R^2	N
−0.075	0.425**	−0.132	−0.511**	—	0.94	60
(0.54)	(2.45)	(1.05)	(3.10)			
−0.105	0.426**	—	—	0.408**	0.94	60
(0.75)	(2.42)			(2.61)		
−0.0817	0.425**	−0.112	−0.510**	—	0.95	60
(0.58)	(2.43)	(0.84)	(3.08)			
−0.113	0.421**	0.385	−0.271	—	0.94	60
(0.79)	(2.35)	(3.11)	(1.57)			
−0.136	0.397*	—	—	0.128	0.92	60
(0.82)	(1.91)			(0.74)		

1997. They indicated the wholesaler's cost as paid to the manufacturer; (b) the HMO price was that paid in the second half of 1997 by a leading California health maintenance organization.

4. These prices were converted into the price for a 'usual daily dose' as determined by *The Medical Letter*, **39**, 11 April 1997.
5. Drug generations were provided in Charles F. Lacy et al., *Drug Information Handbook*, 6th edn, 1998–99, Lexi-Comp Clinical Reference Library, pp. 1391–2.

generic drugs than do pharmacies. Again brands facing generic competition appear to be priced higher than in those markets where generic entry has not yet occurred.[10] Note that equations 3 and 4 duplicate the earlier ones, except that the attribute variables are dichotomous.

As in the previous case, we tested whether the HMO/pharmacy coefficients were different and found, again, that they were not. Even the pharmacy branded product variable is not significantly greater than the corresponding HMO variable for this class of drugs. We also examined the residuals from the first equation and tested for heteroscedasticity. As before, we found its presence; but again there were not substantive differences in the regression results after accounting for this problem, except for minor changes in the recalculated *t* statistics.

Antimicrobials

The empirical findings for antimicrobial drugs are shown in Table 4.3. The first set of coefficients describe the estimated effects of the three efficacy measures on price: the ability to fight streptococcus B, upper

Table 4.2 Calcium channel blockers: factors determining 1997 prices paid by HMOs and pharmacies (dependent variable: natural log of price paid for a usual daily dose)

	Intercept	Side-effect profile	Convenience	Branded product
1.	−3.201**	−0.302	0.487*	1.108**
	(2.65)	(1.26)	(2.29)	(4.16)
2.	−2.494*	−0.556**	0.492*	1.067*
	(2.26)	(3.56)	(2.21)	(1.80)
3.	−1.511**	−0.687*	0.531**	1.063**
	(4.26)	(2.09)	(2.39)	(3.98)
4.	−1.361**	−0.955**	0.599**	0.919**
	(6.42)	(5.02)	(2.75)	(3.44)

Notes on calcium channel blockers equations:

1. Therapeutic properties, side-effect profiles and convenience aspects were determined with the assistance of Neil Wenger, MD.
2. All values were provided on a five-point scale. In equations 3 and 4, cells were combined to create a more conventional two-point scale.
3. Price data were obtained from two sources: (a) pharmacy prices were derived from the price lists distributed by a major drug wholesaler and applicable to the fourth quarter of 1997. They indicated the wholesaler's cost as paid to the manufacturer; (b) the HMO

respiratory infections and urinary tract infections. The coefficient for the streptococcus B variable is significant, with the expected sign. The other therapeutic variables are less significant. There is also little variation in the side-effect variable, leading to an insignificant coefficient for this variable. The coefficient for convenience is positive, however, and significant across all specifications; the estimated coefficient in equation 2 implies that an improvement by one unit leads to an increased average price of $3.56 per day. On the other hand, while the HMO brand and generic effects are not significant, they are negative, indicating that HMOs pay less for both branded and generic products for this class of drugs. Newer antimicrobials also command higher prices, which may be due to their increased efficacy.

One difficulty with these equations is the wide variety of patient conditions these drugs are used to treat. Prices reflect the demand for these drugs by all consumers, not just patients with particular illnesses. Again we found no significant differences between the product attribute and pricing coefficients, and also heteroscedasticity in the residuals from the first equation although, again, this latter finding had no qualitative importance.

HMO brand	HMO generic	Second generation	Third generation	Generic competition	R^2	N
−0.091	0.948*	0.817**	0.295	—	0.71	46
(0.39)	(2.08)	(2.73)	(0.83)			
−0.059	1.018**	—	—	0.447*	0.68	46
(0.25)	(2.78)			(2.09)		
−0.139	0.769*	0.596*	0.064	—	0.72	46
(0.62)	(2.22)	(1.81)	(0.18)			
−0.158	0.763*	—	—	0.548**	0.73	46
(0.73)	(2.27)			(2.73)		

price was that paid in the second half of 1997 by a leading California health maintenance organization.

4. These prices were converted into the price for a 'usual daily dose' as determined by *The Medical Letter*, **38**, 16 February 1996.
5. Drug generations were determined by introduction dates: first generation, all products launched prior to 1982; second generation, all products launched between 1982 and 1990; third generation, all products launched after 1990.

Antihistamines

Table 4.4 shows the results for the antihistamine therapeutic group. The therapeutic effectiveness variable in our hedonic regression is significant when the variable is dichotomous but not when a five-point scale is used as in equations 5 and 6. Side-effects detract from price, but the coefficients are not significant. Convenience, however, appears as a significant price determinant for this class of drugs, with the estimated coefficient from equation 3 indicating that price increases on average by \$3.38 per day, with a one unit improvement. Brand effects are strongly positive, but other variables are not consistently significant, including the HMO prices paid for generics, the generation of the drug, or whether or not there is generic competition. The small sample size of drugs makes it difficult to estimate coefficients with precision. As before, we find heteroscedastic residuals, although there was no substantive importance for these findings.

Ophthalmic Solutions for Glaucoma

There is little variation in the therapeutic effectiveness of ophthalmic solutions used to treat glaucoma,[11] although these drugs tend to vary

Table 4.3 Antimicrobials: factors determining 1997 prices paid by HMOs and pharmacies (dependent variable: natural log of price paid for a usual daily dose)

	Intercept	Strep B	URI	UTI	Side-effect profile	Conven.	Branded product
1.	−6.222*	0.805**	−0.934	0.228	0.423	1.543**	0.232
	(2.37)	(2.96)	(1.21)	(0.85)	(0.71)	(3.06)	(0.15)
2.	−7.767**	0.570**	—	—	0.239	1.271**	1.005*
	(4.08)	(3.17)			(0.53)	(3.71)	(1.99)
3.	−7.111**	0.463**	—	—	0.411	1.072**	1.756**
	(3.62)	(2.60)			(0.86)	(3.29)	(2.53)
4.	−3.965**	1.807**	1.526	1.675	−1.430*	2.627**	1.288
	(7.34)	(4.12)	(1.52)	(1.62)	(1.87)	(5.14)	(1.00)
5.	−3.633**	1.570**	—	—	−0.842	2.518**	0.900*
	(7.58)	(4.07)			(1.34)	(5.83)	(2.08)
6.	−3.616**	1.520**	—	—	−0.791	2.555**	2.20**
	(7.37)	(3.85)			(1.23)	(5.79)	(4.58)

Notes on antimicrobial equations:

1. Therapeutic properties, side-effect profiles and convenience aspects were determined with the assistance of Neil Wenger, MD. Strep B, URI and UTI variables indicate, respectively, a drug's effectiveness in treating streptococcus B, upper respiratory infections and urinary tract infections.
2. All values were provided on a five-point scale; in equations 4, 5 and 6, cells were combined to create a more conventional two-point scale.
3. Price data were obtained from two sources: (a) pharmacy prices were derived from the

substantially in the side-effects associated with treatment (see Table 4.5). Average prices increase by $1.69 per day, from equation 2, when side-effects are more serious by one unit. Although the coefficients of the side-effect and brand variables are significant with the anticipated signs, the coefficient for generic competition has the opposite sign from what is expected. HMO prices do not appear to differ from pharmacy prices for these drugs. The generation of these products is an important determinant of price, although the direction of effect is unexpected, with both second- and third-generation drugs tending to be priced significantly *lower* than first-generation drugs. For these drugs as well, we found similar attribute and pricing coefficients along with heteroscedasticity, but again there were no substantial differences in the regression results due to the latter problem.

HMO brand	HMO generic	Second generation	Third generation	Generic competition	R^2	*N*
−0.410	−0.886	1.566	0.379	0.791	0.68	50
(0.86)	(1.45)	(1.24)	(0.25)	(0.51)		
−0.406	−0.923	1.347*	0.100	—	0.67	50
(0.88)	(1.55)	(2.05)	(0.12)			
−0.292	−0.904	—	—	−0.705	0.63	50
(0.61)	(1.47)			(1.05)		
−0.403	−1.075*	1.723	1.170	−09.370	0.77	50
(1.00)	(2.09)	(1.62)	(0.95)	(0.30)		
−0.380	−1.075*	1.418**	1.378**	—	0.76	50
(0.97)	(2.11)	(2.72)	(2.63)			
−0.300	−1.102*	—	—	−1.239**	0.74	50
(0.75)	(2.13)			(2.66)		

price lists distributed by a major drug wholesaler and applicable to the fourth quarter of 1997. They indicated the wholesaler's cost as paid to the manufacturer; (b) the HMO price was that paid in the second half of 1997 by a leading California health maintenance organization.

4. These prices were converted into the price for a 'usual daily dose' as determined by *The Medical Letter*, **38**, 16 February 1996.
5. Drug generations were determined by introduction dates: first generation, all products launched prior to 1982; second generation, all products launched between 1982 and 1990; third generation, all products launched after 1990.

9 CONCLUSIONS

In this study, we estimate hedonic price equations for drugs in five important therapeutic classes. Our price equations included product attributes such as therapeutic benefits, side-effect profiles and convenience factors, and also market variables including whether the purchaser was an HMO or not, as well as the existence of generic versions of the drug.

Although our results differed somewhat across therapeutic category, the equations generally demonstrated a highly significant explanatory power, with the R^2 in each class as high as 0.72, and often exceeding 0.90. The results indicate that more effective drugs and those with fewer side-effects and greater convenience command higher prices. We also confirm our expectations that branded products have much higher prices than their generic counterparts, and also that generic competition tends to increase branded prices.

Table 4.4 Antihistamines: factors determining 1997 prices paid by HMOs and pharmacies (dependent variable: natural log of price paid for a usual daily dose)

	Intercept	Therapeutic benefit	Side-effect profile	Convenience	Branded product
1.	−6.708**	0.415	−0.261	1.092**	1.441**
	(2.81)	(1.27)	(1.40)	(2.62)	(5.72)
2.	−5.499**	0.195	−0.222	0.988**	0.788*
	(2.35)	(0.56)	(1.25)	(3.12)	(2.14)
3.	−6.807**	0.406	−0.254	1.127**	1.445**
	(3.04)	(1.29)	(1.44)	(3.56)	(5.85)
4.	−1.953*	0.083	−0.792	0.408	1.430
	(2.35)	(0.13)	(1.41)	(0.69)	(5.62)
5.	−1.162*	−0.838**	−0.668	0.652	1.121**
	(1.80)	(2.76)	(1.17)	(1.11)	(3.31)
6.	−1.152*	−0.776**	−0.735	0.586	1.489**
	(1.77)	(2.56)	(1.29)	(0.99)	(5.81)

Notes on antihistamine equations:
1. Therapeutic properties, side-effect profiles and convenience aspects were determined with the assistance of Neil Wenger, MD.
2. All values were provided on a five-point scale; in equations 4, 5 and 6, cells were combined to create a more conventional two-point scale.
3. Price data were obtained from two sources: (a) pharmacy prices were derived from the price lists distributed by a major drug wholesaler and applicable to the fourth quarter of

The findings differ somewhat across therapeutic categories, with product attributes having a greater influence on price in some classes than others. An interesting result is that the value placed on more convenient dosing forms is relatively similar in antimicrobials and antihistamines, but much lower in calcium channel blockers.

Some important results from this study concern the prices paid by HMOs relative to those charged to pharmacies. There has been considerable comment that HMOs pay less for branded products than do pharmacies, and there is support here for that result. However, there has not been much discussion of the relative prices paid for generic products. On this issue, we find that HMOs may pay *more* for generics. This observation requires more attention than it has received, and particularly more explanation as to its underlying cause.

We also tested whether it was appropriate to pool the HMO and pharmacy data. In all cases, we could not reject the hypothesis that product attributes were valued the same by these two types of buyers.

HMO brand	HMO generic	Second generation	Third generation	Generic competition	R^2	N
0.600*	0.179	0.078	−0.578	—	0.90	43
(1.71)	(0.65)	(0.13)	(0.90)			
0.461	0.165	—	—	0.740*	0.90	43
(1.32)	(0.61)			(1.92)		
0.602*	0.176	—	−0.649*	—	0.90	43
(1.74)	(0.65)		(1.78)			
0.587	0.183	1.078	0.377	—	0.90	43
(1.61)	(0.66)	(1.51)	(0.64)			
0.401	0.168	—	—	0.439	0.89	43
(1.11)	(0.60)			(1.26)		
0.473	0.178	—	−0.336	—	0.89	43
(1.31)	(0.63)		(0.92)			

1997. They indicated the wholesaler's cost as paid to the manufacturer; (b) the HMO price was that paid in the second half of 1997 by a leading California health maintenance organization.

4. These prices were converted into the price for a 'usual daily dose' as determined by *The Medical Letter*, **38**, 16 February 1996.
5. Drug generations were determined by introduction dates: first generation, all products launched prior to 1982; second generation, all products launched between 1982 and 1990; third generation, all products launched after 1990.

The effect of product generations is not consistent across drug category, with newer generations having higher prices in the cases of calcium channel blockers and antimicrobials, but lower prices for antidepressants and glaucoma drugs. This result is hardly surprising since newer generations of drugs may not represent important therapeutic advances in every class. Where they do, quality-adjusted prices are likely to be higher, but not otherwise.

The hedonic approach, expanded to account for the unique market characteristics of pharmaceuticals, can be useful for identifying the primary factors affecting pharmaceutical prices. Since these prices are often buyer-specific, in that different buyers may pay different prices for the same product, we have moved beyond earlier studies that emphasized biochemical attributes, and have also considered the impact of buyer characteristics. This procedure can be used to determine whether specific prices are higher or lower than those predicted by the product attributes alone. For that purpose, however, a larger study, including more therapeutic drug

Table 4.5 Glaucoma opthalmic solutions: factors determining 1997 prices paid by HMOs and pharmacies (dependent variable: natural log of price paid for a usual daily dose)

	Intercept	Side-effect profile	Convenience	Branded product
1.	−0.433	−0.462**	−0.053	1.564**
	(0.59)	(3.29)	(0.59)	(6.50)
2.	−0.379	−0.526**	0.030	1.169**
	(0.48)	(3.47)	(0.030)	(5.23)
3.	−2.218**	−0.205	0.146*	1.227**
	(3.48)	(1.45)	(1.73)	(5.39)
4.	−1.204	−0.347**	0.030	1.235**
	(1.76)	(2.49)	(0.34)	(5.87)
5.*	−1.416**	−0.605**	−0.105	1.688**
	(4.72)	(3.19)	(0.62)	(7.01)
6.*	−1.327**	−0.735**	−0.144	1.302**
	(4.02)	(3.58)	(0.76)	(5.61)
7.*	−2.33**	−0.152**	0.137	1.364**
	(8.784)	(0.79)	(0.71)	(5.63)
8.*	−1.821**	−0.415*	−0.028	1.330**
	(6.48)	(2.19)	(0.16)	(6.10)

Notes on glaucoma equations:

1. Therapeutic properties, side-effect profiles and convenience aspects were determined with the assistance of Patrick Tso, MD.
2. All values were provided on a five-point scale. In equations 4–8, cells were combined to create a more conventional two-point scale.
3. Price data were obtained from two sources: (a) pharmacy prices were derived from the price lists distributed by a major drug wholesaler and applicable to the fourth quarter of 1997. They indicated the wholesaler's cost as paid to the manufacturer; (b) the HMO

classes, a larger sample of prices and a more rigorous approach to determining drug characteristics, would be needed.

These findings demonstrate that product attributes, including efficacy, safety and convenience, are valued by drug purchasers. The increased willingness to pay for drugs that incorporate these features provides an additional incentive for pharmaceutical manufacturers to develop products that include these characteristics. The implications of this study for industrial policy are clear. Drug attributes such as efficacy, safety and convenience tend to be valued by consumers and their agents. Administered pricing programmes that do not recognize consumer value from all of these dimensions will reduce incentives for research efforts on product attributes that consumers value. In other words, prices do influence innovation.

HMO brand	HMO generic	Second generation	Third generation	Generic competition	R^2	N
−0.131	0.278	0.847**	−0.670**	−0.688**	0.70	53
(0.71)	(1.18)	(3.93)	(2.45)	(3.12)		
−01.41	0.254	−0.681**	−0.377	—	0.64	53
(0.69)	(0.98)	(2.98)	(1.35)			
−0.110	0.176	—	—	−0.492*	0.72	53
(0.52)	(0.66)			(2.09)		
−0.109	0.235	−0.483**	—	−0.503*	0.66	53
(0.55)	(0.94)	(2.92)		(2.30)		
−0.134	0.241	−0.998**	0.787**	0.739**	0.68	53
(0.70)	(0.98)	(4.61)	(2.78)	(3.29)		
−0.150	0.214	−0.897**	−0.535*	—	0.60	53
(0.71)	(0.79)	(3.80)	(1.78)			
−0.117	0.194	—	—	−0.621**	0.52	53
(0.51)	(0.66)			(2.41)		
−0.110	0.224	−0.586**	—	−0.570**	0.62	53
(0.53)	(0.86)	(3.47)		(2.46)		

price was that paid in the second half of 1997 by a leading California health maintenance organization.

4. These prices were converted into the price for a 'usual daily dose' as determined by *The Medical Letter*, **39**, 6 June 1997.
5. Drug generations were determined by introduction dates: first generation, all products launched prior to 1982; second generation, all products launched between 1982 and 1990; third generation, all products launched after 1990.

NOTES

* The authors appreciate the comments and suggestions of Michael D. Intriligator (University of California Los Angeles), who was particularly helpful with the empirical methodology used in this chapter.

1. See also DiMasi et al. (1991), OTA (1993) and Comanor (1999).
2. In recent years, production costs of companies primarily producing branded products have varied between 28 and 35 per cent of total sales. See Comanor and Schweitzer (1995, p.179).
3. Patent protection is eroded by competition from products that are similar in effect, although not chemically identical.
4. See Frank and Salkever (1997), Reiffen and Ward (2002).
5. To be sure, differences in demand elasticities depend on the availability of therapeutic substitutes. These substitutes, however, are present in the therapeutic categories considered here.

6. An example of this price disparity is that reported for Cephalosporins, which is a sub-class of antibiotics. Between October 1985 and August 1996, price increases on sales to pharmacies increased on average by 4.1 per cent per annum, while prices declined on average by 1.4 per cent per annum on sales to hospitals (Ellison and Hellerstein, 1999, pp. 126–7).
7. Products are here viewed as bundles of various characteristics on which final prices depend. See Griliches (1961, 1971), Lancaster (1971), Cowling and Cubbin (1971), Triplett (1969).
8. Earlier studies that applied hedonic methods to pharmaceuticals include Suslow (1996), Berndt et al. (1996) and Cockburn and Anis (1998). While these studies are important precursors to the current one, none of them distinguished between the prices paid for the same drug by different buyers, and they therefore ignore an essential element of pharmaceutical markets.
9. See Feenstra (1995).
10. This result is explained in Frank and Salkever (1997).
11. Curiously, when a therapeutic benefit variable was included, its coefficient exhibited a negative sign.

REFERENCES

Abramowicz, M. (ed.), *The Medical Letter*, **38**, 16 February 1996, **39**, 11 April 1997, New Rochelle, NY.

Berkow, R. (ed.), *The Merck Manual*, 16th edn, Merck Research Laboratories, 1992.

Berndt, Ernst R., Iain M. Cockburn and Avi Griliches, 'Pharmaceutical innovations and market dynamics, tracking effects on price indexes for antidepressant drugs', *Brookings Papers: Microeconomics*, 1996, 133–88.

Cockburn, Iain M. and Aslam H. Anis, 'Hedonic analysis of arthritis drugs', working paper 6574, National Bureau of Economic Research, May 1998.

Comanor, W.S., 'The political economy of the pharmaceutical industry', *Journal of Economic Literature*, **24**, September 1986, 1178–217.

Comanor, W.S., 'The pharmaceutical research and development process and its costs', *Drugs for Communicable Diseases: Stimulating Development and Securing Availability*, Médecins Sans Frontières, Paris, 1999.

Comanor, W.S. and S.O. Schweitzer, 'Pharmaceuticals', in Walter Adams and James Brock (eds), *The Structure of American Industry*, 9th edn, 1995, pp. 177–96.

Congressional Budget Office, 'How increased competition from generic drugs has affected prices and returns in the pharmaceutical industry', July 1998, available at www.cbo.gov.

Cowling, K. and J. Cubbin, 'Price, quality and advertising competition: an econometric investigation of the United Kingdom car market', *Economica*, **38**, November 1971, 378–94.

Dimasi, J., R.W. Hansen, H.G. Giabowski and L. Lasagna, 'The cost of innovation in the pharmaceutical Industry', *Journal of Health Economics*, **10**, 1991, 107–42.

Ellison, Sara Fisher and Judith K. Hellerstein, 'The economics of antibiotics: an exploratory study', in Jack E. Triplett (ed.), *Measuring the Prices of Medical Treatments*, Washington: Brookings Press, 1999, pp. 118–43.

Epple, Dennis, 'Hedonic prices and implicit markets: estimating demand and supply functions for differentiated products', *Journal of Political Economy*, February 1987, **95**, 59–80.

Feenstra, Robert C., 'Exact hedonic price indexes', *Review of Economics and Statistics*, **77**, November 1995, 634–53.

Frank, R. and David Salkever, 'Generic entry and the pricing of pharmaceuticals', *Journal of Economics and Management Strategy*, **6**, 1997, 75–90.

Griliches, Z., 'Hedonic price indexes for automobiles: an econometric analysis of quality change', *The Price Statistics of the Federal Government* (general series no. 73), New York: National Bureau of Economic Research, 1961, pp. 137–96.

Griliches, Z., 'Introduction: hedonic price indexes revisited', *Price Indexes and Quality Change: Studies in New Methods and Measurement*, Cambridge: Harvard University Press, 1971, pp. 3–15.

Kettler, Hannah E., *Updating the Cost of a New Chemical Entity*, London: Office of Health Economics, 1999.

Kmenta, Jan, *Elements of Econometrics*, London: Macmillan, 1986.

Lacy, Charles et al., *Drug Information Handbook*, 6th edn, Hudson, OH: Lexi-Comp, Inc., 1998–99.

Lancaster, K.J., *Consumer Demand: A New Approach*, New York: Columbia University Press, 1971.

Lu, Z. John and William S. Comanor, 'Strategic pricing of new pharmaceuticals', *Review of Economics and Statistics*, February 1998, 108–18.

Reiffen, David and Michael R. Ward, 'Generic Drug Industry Dynamics', working paper, US Treasury Department, October 2002.

Suslow, Valarie Y., 'Measuring quality change in the market for anti-ulcer drugs', in Robert B. Helms (ed.), *Competitive Strategies in the Pharmaceutical Industry*, Washington: American Enterprise Institute, 1996.

Teitelbaum, Fred, 'Drug Trend Report 2002', Express Scripts, Inc., June 2003.

Triplett, J.E., 'Automobiles and hedonic quality measurement', *Journal of Political Economy*, **77**, 1969, 408–17.

US Congress, Office of Technology Assessment, 'Pharmaceutical Research & Development: Costs, Risks, and Reward', February 1993, pp. 50–72.

PART THREE

The Micro View

5. Recent developments in universities regarding intellectual capital and intellectual property*

Emidia Vagnoni, James Guthrie and Peter Steane

1 INTRODUCTION

The western world has entered what is commonly referred to as the 'knowledge age', where information and ideas have overtaken agricultural produce and manufactured goods as key commodities (Dunford et al., 2001). National wealth and economic strength are now being measured in terms of knowledge, its usefulness and the speed with which it can be applied. Nations are being forced to compete in a global information economy where ideas, information and knowledge have no boundaries, instead multiplying and growing at a hectic pace (Petty and Guthrie, 2000). Intellectual capital (IC) and its legal counterpart, intellectual property (IP) are increasingly being seen to have an influence on the overall economy of nations.

Essentially demands of globalization and rapid advances in technology have led to national leaders calling on their nations to become 'knowledge economies'. The very words 'knowledge economy' bring universities, as the producers and transmitters of knowledge, to the forefront of the political and public arena. At the same time universities are experiencing cutbacks in government funding, forcing them to find alternative means to meet demand while maintaining their integrity.

Alongside this development, the university sector internationally has also come under pressure to become more 'relevant', to establish commercial partnerships and to collaborate in research endeavours (OECD, 1998). These collaborations can be with other universities, third sector non-profit organizations or private sector profit-making organizations.

In this knowledge-based world, value and the maintenance of a competitive edge depends more and more on the management of ideas and innovation. As governments and their agencies embrace the 'knowledge age', the value of, and demand for, government information and services will increase significantly (OECD, 1999a). It is reasonable to expect that various national, state and regional governments will seek to position themselves in such a way as to be able to manage, develop and use available intellectual assets successfully in order to meet such demands. Given the broader and diffuse nature of such interaction between the public and private sectors (Olson et al., 2001) as well as not-for-profit sectors (Carroll and Steane, 2000), these demands will become more complex and have much wider ramifications. These effects include the need for changing accountability relationships between universities and industries, and increased transparency on outcomes, where, as Martin (2000) states, the opportunities for multiple stakeholders to share in the commercial success of state-funded research are available.

The proactive management of intellectual property (IP) is becoming an increasingly important consideration for government agencies and other bodies in maintaining knowledge results (Barrett, 2001a). Alongside the growing importance of innovation and creativity to national wealth creation (Petty and Guthrie, 2000), an increasing need for detailed research on policy issues associated with the management of university-based intellectual capital (IC), and its legal aspect, intellectual property (IP), can also be perceived.

This chapter reports some preliminary research on the international development of policies regarding university governance, the corporatization of research and the management of university IC and IP. A brief case study is provided which reviews Australian and Italian policy prescriptions concerning innovation, commercial practices and IP protection issues. The aims of the chapter are threefold: (a) to undertake a review of OECD and a number of national government statements concerning the 'knowledge era', the need for innovation and creativity and the role of universities in this process; (b) to identify the current Australian and Italian policy debates concerning the above issues; (c) to use the outcomes of this research to contribute to public policy debate concerning the relationship between universities and industry in different contexts, and to discuss issues concerning the management and exploitation of intellectual capital by business, research institutions and other users of intellectual property systems.

Section 2 outlines key aspects of the knowledge management debate and the rise of IC and IP issues in modern society. The changing role of the public sector, including universities facing this knowledge development will be explored. In section 3, new configurations of university research finance, which have emerged in several countries in response to the fact that universities still fund a large part of basic research activities, but have been struggling for national funds during previous decades, are outlined. Section 4 seeks to discuss both the collaboration and partnerships models that universities are developing as well as some of the implications of these models. Section 5 then outlines the Australian and the Italian university policy issues related to research corporatization and intellectual property. A comparison between these two OECD countries' experiences may be useful to address future directions in the IP issues debate, given the different experience Australia and Italy have in terms of research corporatization. Finally, some issues for future policy debate will be outlined.

2 THE KNOWLEDGE ERA, ORGANIZATIONAL KNOWLEDGE AND INTELLECTUAL PROPERTY

Knowledge may be defined as information integrated with experience and context, subjected to interpretation and reflection. It is a high-value form of information that is ready to apply to decisions and actions. The rising importance of knowledge-intensive employment and the knowledge society is highlighted by various recent OECD reports (OECD, 1998, 1999b, 2001).

Knowledge management is the process of capturing the knowledge gained by individuals and spreading it across the organization, in order to increase the capability of the organization to both create new knowledge and embody it in products, services and systems. Furthermore knowledge management often involves the spreading and transferring of knowledge between organizations for mutual gain. Duffy (2001, p.59) observed: 'the ultimate objective of managing knowledge is to capitalise on the intellectual capital, specifically to encourage knowledge transfer and support knowledge sharing and reuse'. From obscurity, the emphasis on knowledge has come to dominate the corporate mind; 'knowledge workers' is a term that has only existed for some 40-odd years, and today they comprise about a third of the workforce in industrialized nations, a position equal in number to manufacturing workers. The challenge for higher education institutions is to galvanize the knowledge locked within them to provide formal education and skills aimed at expanding the knowledge-acquiring capacity of this rising group within the workforce. At the same time, universities face the challenge of guarding against the dissemination of that knowledge, and the impending threat of its commercialization. Universities will be required less to provide skilled vocationally-oriented graduates, but graduates skilled in their vocational choice. Graduates should have broader skills for acquiring the knowledge necessary to change roles and incorporate differing functions during work: working in teams, managing projects, and so forth.

However, if benefits are to be realized, the right skills, culture and commercial acumen must be put in place, bearing in mind the nature of public service, public university administration and the associated responsibility and accountability to citizens. While explicit knowledge is largely stored in computer systems, a great deal of tacit knowledge resides within people. The major task of management is to deal purposefully with this knowledge. Organizational strategy must include means to identify 'missing' or 'orphaned' knowledge (Caddy et al., 2001) and to be wary of intellectual liabilities (Caddy, 2000).

While perhaps not recognized as such, the management of knowledge has always been at the core of what the public sector and its associated organizations do, and it is inseparable from strategy, planning, consultation and programme implementation and accountability reporting. Similar to the situation with IC, the proactive management of IP is becoming an increasingly important consideration for nations and their agencies, including universities, in maintaining the capabilities of the nation and states to achieve an effective transformation into a networked, collaborative, knowledge-based society (OECD, 2000a).

Barrett (2001a), the auditor-general of the Commonwealth of Australia, defined IP as the rights granted by law in relation to the fruits of human creative activity. His more specific and technical definition is as follows:

> Intellectual Property includes all copyright (including rights in relation to phonograms and broadcasts), all rights in relation to inventions (including patent rights), plant varieties, registered and unregistered trademarks (including service marks), registered designs, circuit layouts, and all other rights resulting from intellectual activity in the industrial, scientific, literary or artistic fields.

Figure 5.1 locates IP in the wider perspective of organizational knowledge (Contractor, 2000). The figure provides clarification of the concepts used, and their interrelationships, particularly when speaking about Intellectual Property (IP), Intellectual Asset (IA), and Intellectual Capital (IC).

Categories (I) and (II) in Figure 5.1 make the distinction between IP that is registered (say, with a patent) and that which is unregistered (for example, drawings), but recognizable in a tangible fashion (codified) and central to an organization's core business, continuity and success. Category (III) is the least tangible (uncodified) asset and is basically tacit intellectual human capital. While the author is concerned to establish a general set of principles for the *valuation* of IP, the main purpose of this chapter is to stress IP as part of a *whole* management of knowledge approach. The issues of measurement, valuation and reporting are not the focus of the current research efforts (see Guthrie, 2001; Guthrie and Petty, 2000; Guthrie et al., 2001 for more details).

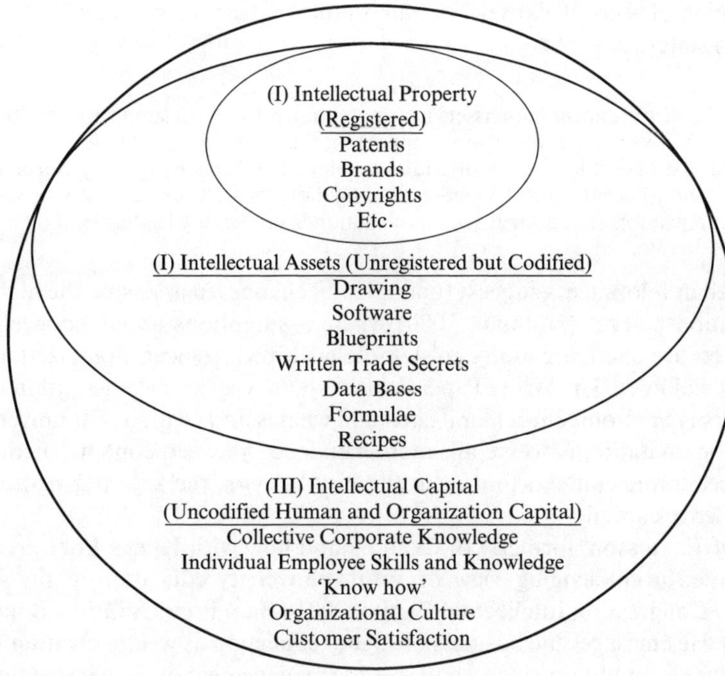

Source: Contractor (2000), Figure 1.

Figure 5.1 Groupings within the organizational knowledge framework

Barrett (2001b) rightly indicates the unpalatable fact that 'many public sector entitles do not know what they own in the form of intangible assets, such as intellectual property' and IC, and the university sector can also be included in this statement.

Intellectual property rights have gained international standing as some of the most important rights that need legal protection. It is not surprising that governments have therefore focused their attention on IP management. For example, the United Kingdom (UK) government is encouraging public bodies to make better use of their IP assets and has focused on the commercial exploitation of the outputs of publicly funded research in a variety of ways, including joint ventures with private sector partners. In 1999, the UK government published a White Paper, 'The Future Management of Crown Copyright' (UK Ministry, 1999) that dealt with the issue of availability and access to government information and government-produced materials. The paper allows government departments freedom to decide how works they originate are to be distributed or commercialized, with the exception of Acts of Parliament and Statutory Instruments. The UKNAO (2001) states:

> public sector information assets have potential, not only in supporting the business of government, but also in supporting the economy as a whole. . . . Where value has been added, or information developed and created within government [. . . and it] is enhanced beyond core obligations or statutory duties to produce that raw information, then such information is potentially tradeable.

Such developments suggest fundamental changes in assessing the higher education market (Holman, 2000) where assumptions about knowledge creation are changing along epistemological, pedagogical, organizational and social lines. The White Paper deals with the way knowledge within the university environment is manifested and what is done with it. For universities, the invitation is to see an interdependence between content, method and economic context, that is, in the last analysis, the key to generating knowledge capital.

Leif Edvinsson, formerly of Skandia and now with Knexa Enterprises, presented a challenging view of future university education at the 4th World Congress on Intellectual Capital 2001 (Hamilton, Ontario) (Bontis, 2001). He envisages the business of higher education as wealth creation via the generation of knowledge, and sees the management of such institutions as a business. Edvinsson lamented the inability of university deans in managing their business to keep pace with the innovation and flexibility evident in industry and technology. While the time, expense and effort afforded to acquire almost any product have brought benefits to the consumer over the past 40 years, similar savings and efficiencies and flexibility are not evident in the acquisition of university degrees.

The above section has outlined several key aspects of the knowledge management debate and identified the importance of IP and IC to modern public services, including universities. Now we can turn our attention to several contemporary policy issues associated with university research.

3 UNIVERSITY RESEARCH IN TRANSITION

Universities and other higher education institutions are key elements in the development of science and other forms of knowledge in OECD countries. They perform basic and applied research, and train researchers and other skilled personnel. Differences, however, exist between OECD countries regarding the position of university research in science and its relationship to research undertaken in other publicly supported research institutions. Several profiles can be drawn on the basis of national cultural background and economic structures. The OECD (1999b, pp. 21–3) offers four profiles for understanding these differences.

In Anglo-Saxon countries, universities are the major source of basic research, but they coexist with public research institutions devoted to sectors of national interest, such as defence, energy, agriculture and medicine. The latter may undertake basic research where needed, although they are generally involved in applied and technical research activities.

In large continental European countries, university research coexists (and cooperates) with a large public sector engaged in basic research in its own laboratories (Germany's Max-Planck Society, France's Centre National de la Recherche Scientifique – CNRS, and Italy's Consiglio Nazionale delle Ricerche – CNR), which are also involved in technical and applied activities, to provide either R&D infrastructures (as in Germany) or mission-oriented activities (as in France and Italy).

In smaller continental European economies, public research tends to be mainly oriented towards technical and industrial research activities, while universities perform most basic research. There are, however, important differences among countries: some have a large public sector (for example, Norway, Iceland and Portugal), while others do not (for example, Switzerland).

Finally, *in East Asian countries*, which were formerly strongly oriented towards technical applications and the assimilation of foreign technology, university research has remained modest, owing to a lack of financial support, overregulation and the burden of teaching tasks. The situation is changing rapidly, however, as these countries, amongst them notably Japan, are boosting their basic research efforts via support for university research.

While university research, including basic research, contributes significantly to innovation and technical change, this contribution is largely indirect. Recently the OECD (1999b) indicated that firms, key actors in innovation, rely little on university (and public) laboratories as a source of information or stimulus for their innovative efforts, as a number of surveys, including recent empirical analyses of national innovation systems, have demonstrated. Moreover spin-offs and business start-ups from university research are a little seen phenomenon. Even in science-based sectors, interaction with competitors, suppliers and customers is considered more important for firms' innovative efforts than information from university or public sector research (OECD/DSTI, 1997b).

Internationally, in response to these issues, there have been significant changes in the past decade in the university environment which have affected the research-related missions of these institutions. In particular, universities are becoming more diverse in structure and more oriented towards national and international economic and industrial needs. At the same time, they have to cope with higher student enrolments, and lower student subsidies, particularly in continental Europe (OECD, 1999b, p. 6). These trends raise serious questions regarding the capacities of universities to continue making a significant contribution to long-term research while maintaining an appropriate balance between research, training, teaching and knowledge transfer.

The recent OECD (1999b) report on university research in transition notes that university research represents between 15 and 35 per cent of overall R&D expenditure in the OECD economies. It also indicates six major implications concerned with university research in the last decade:

1. *Declining government R&D finance*. Government R&D budgets have been reduced in several OECD countries, often leading to a levelling off, or even a decline, in university research support. Traditionally, governments financed 80 per cent or more of university research as a 'public good', but this percentage has been declining, with the result that universities are seeking new sources of financial support.
2. *Changing nature of government finance*. Government funding for academic research is increasingly mission-oriented, contract-based and more dependent on output and performance criteria. This can lead universities to perform more short-term and market-oriented research.
3. *Increasing industry R&D finance*. Private industry is funding an increasing share of research in universities. This support, in the form of joint projects, contract research and financing of researchers, is also leading universities to perform research more directed to potential commercial applications.
4. *Increasing systemic linkages*. The institutional context of research is changing as universities are encouraged to enter into joint ventures and cooperative research with industry, government facilities and other research institutions.
5. *Growing demand for economic relevance*. Universities are under pressure to contribute more directly to the innovation systems of their national economies. However, they are often constrained by rigidities arising from the traditional disciplinary organization of research and other university structures. This can cause considerable tension in the university research environment. It is interesting to note that the OECD report failed to mention the prevalence of corporate support within universities.

6. *Internationalization of university research.* Globalization, stemming partly from advances in information and communications technologies, is affecting the climate for research and the conduct of R&D. It is also making research more competitive and leading to increased specialization.

The OECD (1999b, p. 8) report indicated that universities are recognized as essential to the knowledge-based economy, and that no nation will willingly permit a serious, permanent decline in the research, training or knowledge transfer capabilities of their national systems. In the early part of the 21st century, however, university research and its relation to society are likely to be different from what they were at the end of the 1990s. The report notes that OECD countries need to ensure that universities can continue to perform their functions to the benefit of society at local, national and global levels.

What is needed is university education that encompasses greater flexibility. It is so-called 'knowledge nomads' who are the likely inheritors of the new knowledge economy emerging in innovative universities. These universities are institutions that see their role as brokers of knowledge workers, who are free to generate knowledge wealth, rather than be merely protectionist-style employees of institutions concerned about brand and products. Such defensiveness mimics the enclosed monasteries from which many universities arose and does not deal with the challenges of knowledge creation in today's economies.

One of the fundamental lessons arising from these reformist approaches to university research in the OECD is the importance of flexibility and particularity. There is less interest in centralized and inflexible ways of delivering education services. Knowledge generated through research can be maximized through systemic partnerships. As Figure 5.1 emphasizes, intellectual capital includes the *way* knowledge is managed within organizations and involves some 'creative dislocation', that is, loose networks between researchers and business.

It must be noted that, quantitatively speaking, university research plays a relatively modest role in OECD science systems. In the five largest scientific powers (the United States, Japan, Germany, France and the United Kingdom), it represents about 15 per cent of the total R&D expenditure, and in other countries it ranges between 25 and 30 per cent (OECD, 1999b, p. 20). However, universities do fulfil an essential function as the principal performers of basic research. For the five largest scientific powers, universities undertake 60 per cent or more of basic research. In general, basic research amounts to half of university research, although this share is tending to diminish.

Universities are increasingly involved in applied and technical tasks as their relations with the business sector intensify. At the same time, it should be noted that modern technologies (such as biotechnologies) are increasingly blurring the boundary between basic and applied research (and, to a certain extent, technical development).

In summary, the above discussion indicates the importance of universities in the knowledge-based economy and the changing nature of the financial mix between university and other organizations' R&D expenditure. More importantly it indicates some major implications that must be addressed by universities, particularly the nature of research, research training and the management of knowledge transfer capabilities.

4 COLLABORATION AND PARTNERSHIPS WITH INDUSTRY

With the increasing emphasis on knowledge transfer between the Universities and society, one notable form of knowledge transfer has been the commercialization of research results and innovation. Universities have several stakeholders in this process, including the individual researcher or research team, the private business partners and possibly an entrepreneur who might take the product to market. These relationships require complex contracts regarding ownership of IC and IP, protection of public funds and risk transfer.

The dominant patterns of collaboration between industry and universities relate to technology transfer, with the universities being seen as a source of knowledge to support commercial purposes. A variety of links exist between universities and business enterprises. Knowledge generated by universities can be understood as a public good that needs management in order to maximize the wealth creation activity. Partnerships between universities and industry are part of broader interdependencies between different economic entities. Such linkages can be seen in the rhetoric of Tony Blair's *Third Way* and Gerhard Schroeder's *Neue Mitte*, as well as present efforts by George W. Bush's Office of Faith-Based Action, designed to generate greater alliance between churches and the delivery of government services. In all this interdependency, universities sit at the cusp of an opportunity both to be competitive and to build cohesion at the same time.

There has been a dramatic growth in the development of a very wide range of public–private partnerships. As a crude measure of their extent and importance, a search through GOVBOT, the US federal government's database of web sites using the words 'public–private partnership' resulted in 8830 hits. A similar search through the UK government's CCTA

Government Information Service resulted in over 1000 documents being identified. Influential national lobby groups promoting public–private partnerships have been established in the USA (the National Council for Public Private Partnerships, the successor to the Privatization Council, established in 1985) and Canada (The Canadian Council for Public–Private Partnerships).

The OECD (1998) reports several schemes which have proved effective, and these are described below. The four main schemes are the pre-competitive approach, the centre-based approach, the direct business funding approach and the university entrepreneurial approach. These are now expanded; the first, the pre-competitive approach, involves government funding of university and business research to solve a basic industry problem. For example, Germany has encouraged university–industry partnerships as a means of speeding up technology transfer. German policy towards knowledge transfer has been to support pre-competitive approaches across industries. The OECD (1998) reports that, between 1991 and 1996, some 350 projects were funded in the areas of medicine/pharmacy, food, plant breeding and environmental biotechnology. This was done in an attempt to facilitate the rapid transfer of research results to industry and to increase the R&D activities of small and medium-sized enterprises (SMEs). This support programme was found to have contributed substantially to speeding up the commercialization of biotechnology in Germany.

Collaborative research between different business enterprises and research institutions on a single project has been found to contribute to the better exploitation of limited research capacities through the pooling of resources, the speeding up of technology transfer between science and industry, the generation of synergy and large-scale (as opposed to selective) promotion. As government support is available only at the pre-competitive stage, projects tend to involve basic industrial research. Subsequent company-specific solutions are developed without government support.

The second scheme, centre-based research, seeks to foster cooperation between universities and industry research laboratories. Since the late 1970s, the US National Scientific Fund has had a number of different programmes to facilitate cooperation between university and industrial research laboratories and to promote knowledge transfer. The two most ambitious of these, initiated in the late 1980s, involve the engineering research centres and the science and technology centres, which provided substantial support for up to ten years for research in areas of interest to industry. A recent and highly positive evaluation of the science and technology centres affirmed the value of long-term stable funding and found the centres to have produced research of high scientific quality which could only have been addressed through such centre-based research. The evaluation concluded that dissemination of both their basic and applied research had been highly successful (OECD, 1998). The active cooperation and participation of industry was seen to lead to better research, new ideas, leveraged funding, increased staff appreciation of the industrial sector and better preparation of students entering the workforce.

The third scheme involves direct business funding of university research. Governments are encouraging universities to seek direct industrial funding research support by various means (for example by making support partially conditional on establishing industrial research partnerships). Industrial support, whether in the form of contracts or research partnerships, promotes the integration of universities into the knowledge-based economy. As a result, during the last 15 years, most OECD countries have experienced a higher growth of private funds for research in universities compared to the government funds rate. On average, as seen in Table 5.1 the growth rate of business enterprise funds deployed in the university R&D sector has been 150.8 per cent in the decade 1985–95. In the same period, the growth rate of government funds has been only 25 per cent.

The fourth scheme seeks to encourage universities to establish 'profit' centres and generate their own form of venture capital; that is, to undertake an 'entrepreneurial' approach. By exploiting opportunities for industrial contracts and partnerships, some academic institutions are gradually transforming themselves into partially, or even predominantly, self-financed 'profit' centres. This trend is likely to be amplified, although it will not become the dominant pattern for academic research within national science systems. Concerns expressed in the past about the ownership of intellectual property resulting from university research linked to industry, or about possible limitations imposed on the publication of commercially relevant research results in the open literature, have most often been settled to the satisfaction of both university and industry partners. A more serious problem centres on the fact that not all types of higher education institutions, nor all disciplines, can prosper equally in this way. For example, medical and engineering schools tend to excel in this arrangement; science-oriented institutions, however, find it more difficult, while for most of the social sciences and all of the humanities, it is virtually impossible.

In summary, industrial partnerships can be of benefit to university research. However, there are limits to the financial resources universities can obtain from private industry or other non-government sources.

Table 5.1 Share of total business enterprise funds and government funds in the university R&D sector

Country	Business enterprise funds (%)		Government funds (%)	
	1985	1995	1985	1995
United States	1	1.4	23	31.9
Canada	2.5	4.9	39.2	41.9
Mexico	—	3.6	—	54.3
Japan	0.5	0.7	50.2	47.4
Australia	1.5	1.9	40.5	47.7
New Zealand	—	4.4	—	—
Austria	1.2	1.4	70.7	—
Belgium	2.5	6.7	51.4	—
Czech Republic	—	0.3	—	21.7
Denmark	0.5	0.8	49.3	56
Finland	—	1.9	—	49.6
France	0.7	1	27.4	35.7
Germany	1.3	2.4	34.7	47.3
Greece	—	7.6	28.3	—
Hungary	—	1.1	—	46.4
Iceland	0.7	3.2	35.1	34
Ireland	3	2	33.4	52.4
Italy	0.6	2.5	36.3	44.6
Netherlands	0.5	2.5	50.1	58.1
Norway	2.1	2.8	44.4	53.4
Poland	—	9.6	—	31.4
Portugal	1	1.3	45.5	45.3
Spain	0.5	4.6	42.6	54.2
Sweden	2.5	2.4	66.5	—
Switzerland	0.5	0.7	58.7	—
Turkey	—	29.3	—	89.5
United Kingdom	1.7	2.4	27.8	38.2
Average growth rate (%)	—	150.8	—	25

Source: OECD data (1999b).

While industrial financing can complement government financing, it cannot replace it. While universities can certainly make direct, short-term contributions to national economic goals, substantial core funding unrelated to identifiable short-term goals will continue to be required if universities are to conduct the basic research on which the long-term vitality of the knowledge-based economy depends.

5 CURRENT AUSTRALIAN AND ITALIAN POLICY DEBATE

This section will briefly discuss a number of current university research policy debates in the Australian and Italian context. This brief case study will provide a more in-depth comparison of the two OECD countries' experiences with research corporatization and highlight a number of issues for the future direction of the IC and IP issues debate. We will start with Australia and then turn to Italy.

Australian Policy Debates

In Australia, during the last decade, overall support for university research has been maintained, with a marked increase in contract-based funding and development of output criteria. More funds have been provided for industry–university cooperative schemes such as strategic partnerships with industry research and training (SPIRT).

The management of the commercialization of research is increasingly being seen by many as part of a broader business management imperative cutting across the broad range of university teaching, research and community service functions. To contextualize these statements, the Australian higher education sector (AHES) revenue totalled some Australian $9 600 million in 1999, of which 39 per cent was raised from non-Commonwealth sources. Total higher education sector equity, including commercial entities, was $19 500 million at the end of 1999, compared with $18 800 million when commercial entities are excluded. Some 34 higher education institutions reported having commercial entities in their 1999 financial statements. Nineteen had commercial entities contributing a net surplus (before abnormal items) totalling $43.4 million. Twelve had commercial entities contributing an aggregate new deficit of $8.5 million to their consolidated operating result. For a university, the largest negative effect of commercial entities to the operating result was around $2 million in 1999. The aggregate consolidated revenue for the sector in 1999 increased by $417 million (4.5 per cent) from 1998. Of the $417 million increase, more than a third (around $151 million) was contributed by commercial entities. The contribution of the commercial entities to total sector revenue was 6.1 per cent ($583 million) in 1999.

Over the last few years, many universities have been reviewing the nature of their technology transfer activities (including research, testing and consultancy services, as well as the supply of postgraduate research students). Many have taken action to professionalize their management of commercial activities, activities that, for any business, are inherently risky, demanding that commercial, legal and ethical judgments be made. Measures taken have included separation of functions, clear assignment of responsibilities, appointment of business managers with commercial expertise, formation of boards of directors with a range of commercial experience, promulgation of institutional policies and procedures, and engagement of expertise from outside as required (for example, that of a patent attorney).

Michael Gallagher (2000), head of the Australian Higher Education Division, DETYA, indicated that, in addition to the desirability of documenting the emergence of new policies, structures and procedures being adopted by universities in relation to research commercialization, some questions need to be addressed:

- What new skill sets are required of leadership teams in modern universities and how are those skills obtained?
- How and to what extent do different IP ownership policies motivate researchers to pursue commercialization options?
- What lessons can be learned from the variety of approaches taken to IP scanning?
- What differences (by field of research) exist in terms of conditions and effective strategies for commercialization?
- What is the relationship between effective industry take-up of research and enterprise size?
- What specific lessons have been learned from experiences with incubators and science/technology parks?
- What are the roles of the humanities and social sciences in innovation?

Current thinking within the IMHE group of the OECD (OECD, 2000a) suggests that large firms look to university research to complement their own R&D in, for example, chemistry. However, in biosciences, IT, pharmaceuticals and computer graphics, where the distinction between basic and applied science has largely dissolved, scientific parity rather than division of labour exists between industry and academic workers. Research-based relationships with small entrepreneurial firms or start-ups generally pose greater complexities for universities than their contacts with large firms. Small firms, lacking the financial resources to support overheads for internal R&D, focus on applying research findings to developing marketable products and seek direct involvement of university scientists and, lacking cash, pay with equity. It is in this area that conflicts of interest can arise for academic researchers. This is also the area of greatest involvement in Australia, fostered through the creation of research parks and business incubators.

A recent report of the OECD's (2001) working group on national innovation systems stresses that 'innovation patterns are highly country-specific, depending on the individual country's economic specialisation and institutional set-up'. It would therefore be useful to know much more about Australian circumstances, including the following:

- In what significant ways does Australia's industry and enterprise structure differ from US and European structures, and what do these differences mean for the supply of and demand for R&D from industry?
- In what ways does the supply of Australian public sector R&D differ from public sector supply arrangements and concentrations in the USA and Europe?
- What is the extent of industry take-up of Australian research by industry sector, enterprise size and field of research?
- What are the constraints operating on industry take-up of public sector research in Australia by industry sector, field of research and enterprise size?
- What are the main options for commercialization of research outside of industry take-up by either Australian or foreign enterprises?
- What makes for success in business start-ups and spin-offs from university research? What are their expected success rates and how do they compare with those of small businesses generally?

Australian policy makers believe that the best mechanism for knowledge transfer is the mobility of people who have both transferable skills (including research problem-solving abilities) and a good informal network of professional contacts. The cooperative research centres (CRC) programme has supported long-term collaborative ventures linking research and research users from universities, Commonwealth and state-funded research organizations and business enterprises. The programme has also improved Australia's research culture.

Recently Neumann and Guthrie (2001) have provided a detailed case study of the development of academic research policy in Australia from the mid-1980s. Their research highlights a move away from individual academic and disciplinary autonomy in the determination of research, towards a performance economy in research. Through funding policies and strategies, there has been a continual increase in the centralization of ever-scarce resources by the government and a competitive redistribution system that rewards those aspects of research that achieve government policy goals (for example, collaboration and private–public partnerships). Ironically, while the rhetoric of government policies has been that of competition and corporatization, with implied decentralization and freedoms, the impact appears to be stronger government control and intervention (Boyer et al., 1994). International comparisons show that Australian government control of universities and research and publication is above the average of OECD nations.

Current Italian Policy Debate

In Europe, there are notable differences in the treatment of research funding and links with private funds, particularly between the countries with the largest scientific investment. The OECD (1998) report on research activities in different countries highlighted the substantial growth that Italy experienced during the 1970s and 1980s. It also noted that since that time, Italy has seen a regular decline in government support for the R&D sector, the overall university R&D effort being reduced in nominal terms in 1996. Italy spends about 1.4 per cent of GDP on R&D, compared to the OECD average of 2.8.

Universities and other public institutions (such as the National Research Centre,[1] the Italian Space Agency, the National Institute of Nuclear Physics and the National Institute of Health) are required to carry out research. From the mid-1990s, R&D expenditure policy affected an increase in the level of responsibility held by universities over the outcomes of research projects. Public R&D funds (both from the Ministry of Higher Education and from the National Research Centre – CNR) have become heavily focused on research issues identified as priorities by public institutions.

The relatively poor level of national and local research finance should, in theory, be compensated for by means of European Union R&D funding mechanisms. However, in practice, few Italian researchers have been able to obtain such funding, owing to the high level of competition for funding amongst research teams from other European nations.

Recent R&D policy has sought both to concentrate research activities in fewer universities and, within those universities, to concentrate effort in particular research specializations. Thus a number of highly specialized centres have been created and funded (Centri di Eccellenza). This policy reflects in part a response to the increasing levels of competition and growing research costs being experienced by many disciplines. It also reflects the need of scientists in an increasing number of disciplines to use large-scale research facilities which are beyond the resources of any single institution and, in some cases, any single country, to provide. For a number of reasons, however, the trend towards concentration and specialization is likely to be self-limiting.

The University of Ferrara: a case analysis

The University of Ferrara is a medium-sized university in Italy. Since its establishment in the 14th century, its primary research emphasis has been science. Pharmaceutical science, surgery, medicine, biology and embryology account for the bulk of research investment. Partly in recognition of, and partly to enhance, this long tradition, the University of Ferrara was recognized in 2000 as an area of research excellence in the field of infective pathology (Centro di Ėccellenza sullo studio delle patologie infettive), the only such centre in the region. As shown in Table 5.2, the 'health' research area attracts 63 per cent of total research funds, including infrastructure finance, matching the general situation of many other universities.

Amongst the nine research areas (Table 5.2), the majority of funds (75 per cent) are absorbed by the 'health' and 'chemistry' research areas, leaving just 25 per cent of funds for the other seven areas. The research area of Economics receives the least funding (1.9 per cent). This probably reflects the fact that the school was only established recently.

As has been common elsewhere (OECD, 1999b), R&D funding has generated extensive discussion during the last two decades at the University of Ferrara. The traditional approach to research funds management, combined with the rigid institutional structure of Italian research and the recent experience of research training in Italian universities, has resulted in many universities managing research funds through a dedicated non-profit organization, such as the Consorzio Ferrara Ricerche (CFR) at Ferrara University. CFR manages almost 50 per cent of the R&D funds deployed by the University of Ferrara. Of these funds, 75 per cent are supplied by business enterprises, 23 per cent by local authorities, with the final 2 per cent being supplied by government. All government R&D funds are managed through the university's individual departments.

The use of a non-profit organization to manage research funds provides the flexibility necessary to develop a business-oriented management structure for the tertiary R&D sector. The service provided by the CFR benefits not only local enterprises: it performs the bulk of its services for enterprises located in other regions of the country (69 per cent) and also abroad (10 per cent). The business enterprise funds undertaken at the University of Ferrara (through the CFR) support both the 'evergreen project' and applied research, in addition to basic research.

From the late 1990s, the research policy has emphasized the link between outcome, output and results of the research. Following the new national R&D policy, the University of Ferrara has shown an increased willingness to undertake collaborative research projects. As a result, a more business-oriented attitude has been developed. Researchers believe that 44 per cent of research outcomes could generate patents, among which 32 per cent

Table 5.2 University of Ferrara, share of business enterprise funds and government funds in the university R&D areas

Research areas	Private funds	%	Public funds (ex 40%, ex 60%, CNR, UE, public organizations)*	%	Total research funds	%
HEALTH (biology, medicine, embryology, surgery, pharmaceutical science)	13915	72.3	7368	51.1	21283	63.2
CHEMISTRY (chemistry, cosmetics, biochemistry)	2473	12.8	1487	10.3	3960	11.8
ARCHITECTURE	331	1.7	498	3.4	829	2.5
ENGINEERING	915	4.8	998	6.9	1913	5.7
BIOMEDICINE	319	1.7	693	4.8	1012	3
ENVIRONMENT SCIENCE (agriculture, geology, botanics)	439	2.3	765	5.3	1204	3.6
ECONOMICS	240	1.2	410	2.8	650	1.9
PHYSICS (physics and maths)	259	1.3	1432	10	1691	5
HUMAN SCIENCE and LAW	0	0	775	5.4	775	2.3
WORKSHOP	360	1.9	—	—	360	—
Total	19251	100	14426	100	33677	100

Note: * This column synthesizes the different kinds of governmental research funds the university researchers can apply for. Thus national research funds managed by the Ministry of High Education (ex-40%), local research funds managed by University of Ferrara (ex-60%), National Centre for Scientific Research funds (CNR), European Union funds (EU) and other public sector organizations funds are considered.

Source: University of Ferrara data and Consorzio Ferrara Ricerche data.

could be European patents. The corporatization of research has thus raised the issue of intellectual property. As can be seen from Table 5.2, the departments of pharmaceutical science, biology and medicine are undertaking a strong programme of corporatization. Within this new collaborative approach, the business enterprise bears the risk of failure and the potential benefits of a successful new product launch.

Another aspect of the corporatization of research requiring attention concerns spin-offs from research outcomes. The Ministry of Higher Education (2000) reports that it has noted a low incidence of such spin-offs. This is partly due to the weak flow of patents and the bureaucracy of the Italian funding system and demonstrates the weak link between university R&D and market business. In an attempt to remedy this problem, the University of Ferrara spin-off centre has been established. In less than one year, it has already generated business valued at $US400 000.

6 SUMMARY AND FUTURE POLICY ISSUES

The higher education sector has performed an increasing share of government-funded R&D in most countries. The relative decline of government financial support has been partly compensated by support from the business sector, which nonetheless remains relatively modest in the vast majority of countries. Moreover overall business R&D efforts have also been severely affected by the economic recession characterizing many OECD countries during the last decade.

These trends are relatively worrying, in that university research will continue to depend largely on government funding, even if funding from other sources necessarily grows. In fact, support for university research will likely be determined by the importance given to the R&D budget in overall government budgets.

Government policies intended to link academic research more directly to other sectors of the economy, and to require that at least some of the research supported by public funds be performance-based, are consistent with the importance of academic research to the knowledge-based economy. However, if carried to extremes, they can distort and undermine academic research by obliging universities to focus excessively on short-term research that could be carried out in other types of institution. This may be detrimental to the traditional mission of universities to conduct long-term, curiosity-driven research and to impart knowledge to a new generation of students.

NOTES

* The authors are indebted to the comments and critiques provided by colleagues attending the last Ferrara Health Industry Policy Forum. The authors would also like to express their appreciation to research assistant John Scott of Macquarie Graduate School of Management. Furthermore we appreciate the support given by the Ferrara University via research grants, which made this project possible. We would like to thank the organizers

and editors of the book, Stuart Schweitzer and Marco Di Tommaso, for their wonderful support and encouragement during this project.

1. Besides funding research grants and scholarships for research training, and advising government, the CNR itself undertakes scientific work directly in 289 research bodies. Of these, 115 study centres and 17 research groups involve collaboration with universities and other agencies.

REFERENCES

Barrett, P., 'Managing and accounting for intellectual property in the public sector', Auditor-General for Australia, address to Australian Government Solicitor Seminar, Identify, Protect and Defend your Intellectual Property Assets, Canberra, 24 July, 2001a.

Barrett, P., 'Retention of corporate memory and skills in the public service – more than survival in the new millennium', address to the 2001 Australasian Council of Public Accounts Committees, 6th biennial conference, 6 February, 2001b.

Bontis, N., 'Thought leadership on intellectual capital', *Journal of Intellectual Capital*, **2**(3), 2001, 183–91.

Boyer, E.L., Altbach, P.G. and Whitelaw, M.J., *The Academic Profession: An International Perspective*, Princeton, NJ: The Carnegie Foundation for the Advancement of Teaching, 1994.

Caddy, I., 'Intellectual assets and liabilities', *Journal of Intellectual Capital*, **2**(1), 2000, 129–46.

Caddy, I., Petty, R. and Guthrie, J., 'Orphan knowledge potential: an exploration of current Australasian practice in the management of intellectual capital', *Journal of Intellectual Capital*, **2**(4), 2001, 384–97.

Carroll, P. and Steane, P., 'Public–private partnerships: sectoral perspectives', in S. Osborne (ed.), *Public–Private Partnerships for Public Services: An International Perspective*, London: Routledge, 2000.

Contractor, F.J., 'Valuing corporate knowledge and intangible assets: some general principles', *Journal of Knowledge and Process Management*, **7**(4), October–November 2000, 245.

Duffy, J., 'Managing intellectual capital', *The Information Management Journal*, **35**(2), April 2001, 59.

Dunford, R., Steane, P. and Guthrie, J., 'Introduction: intellectual capital, the management of knowledge and organizational learning', *Journal of Intellectual Capital*, **2**(4), 2001, 339–43.

Gallagher, M., 'New directions in australian research and research training policy – some questions for researchers', the Annual Conference of The Australian Network for Higher Education Policy Research, Australian National University, Canberra, 7–8 December 2000.

Guthrie, J., 'The Management, measurement and the reporting of intellectual capital', *Journal of Intellectual Capital*, **2**(1), 2001, 27–41.

Guthrie, J. and Petty, R., 'Intellectual capital: Australian annual reporting practices', *Journal of Intellectual Capital*, **1**(3), August 2000, 241–51.

Guthrie, J., Petty, R. and Johanson, U., 'Sunrise in the knowledge economy: managing, measuring and reporting intellectual capital', *Accounting, Auditing and Accountability Journal*, **14**(4), 2001, 365–82.

Holman, D., 'Contemporary models of management education in the UK', *Management Learning*, **31**(2), 2000, 197–217.
Martin, M., *Managing University–Industry Relations*, Paris: International Institute for Educational Planning (IIEP), 2000.
Ministero dell'Università e della Ricerca Scientifica e Tecnologica (MURST), 'Linee Guida del Programma Nazionale di Ricerca' proposte del MURST, accolte dal C.I.P.E. nella seduta del 25.05.2000 e recepite nel D.P.E.F. approvato dal Consiglio dei Ministri il 29.06.2000.
Neumann, R. and Guthrie, J., 'The corporatisation of research in Australian higher education', CPAJ Conference on Corporatisation in Universities, Symposium 2001 on 'The university in the new corporate world', University of South Australia, 18 July 2001.
OECD, *University Research in Transition: Country Notes*, Publication Service, Paris, OECD, 1998.
OECD, 'Guidelines and instructions for OECD symposium', International Symposium Measuring and Reporting Intellectual Capital: Experiences, Issues and Prospects, June, Amsterdam, OECD, 1999a.
OECD, *Science Technology Industry. University Research in Transition*, Publication Service, Paris: OECD, 1999b.
OECD, 'Research management at the institutional level', report of Expert Meeting of the IMHE, Paris, 8–9 June 2000a.
OECD, 'Focus groups on national innovation systems: draft final report: knowledge markets and innovation systems, nurturing the institutions of innovation', working group on innovation and technology policy, DSTI, Paris, November, 2000b.
OECD, 'The new economy: beyond the hype', final report on the OECD Growth Project, meeting of the OECD Council at ministerial level, Paris, 2001, p. 12.
OECD/DSTI, 'The evaluation of scientific research: selected experiences', OECD/GD (97)194, Paris, 1997a.
OECD/DSTI, *National Innovation Systems*, Paris: Brochure, 1997b.
Olson O., Humphrey, C. and Guthrie, J., 'Caught in an evaluatory trap: the dilemma of public services under NPFM', *European Accounting Review*, **10**(3), 2001, 505–22.
Petty, R. and Guthrie, J., 'Intellectual capital literature review: measurement, reporting and management', *Journal of Intellectual Capital*, **1**(2), June 2000, 155–76.
UK Ministry for the Cabinet Office, *The Future Management of Crown Copyright*, White Paper, London: Her Majesty's Stationery Office, March 1999.
United Kingdom National Audit Office (UKNAO), 'Commercialization projects in the United Kingdom and their audit', Paper presented to the INTOSAI Working Group on the Audit of Privatisation, 8th meeting, Budapest, 11–12 June 2001, para. 12.

6. Intangible assets in the European health industry: the case of the pharmaceutical sector

Patrizio Bianchi and Sandrine Labory

1 INTRODUCTION

A report to the European Commission (Gambardella et al., 2000) argues that the European drug industry is losing competitiveness with the USA, although there are differences among EU countries. In particular, the EU is lagging behind in its ability to generate, organize and sustain innovative processes and appears less able to translate R&D into commercial success, partly owing to a strategy of reliance on external inputs such as licences from international companies, pricing policies or peculiarities of the public regulatory and health care systems, rather than a strategy of reliance on own R&D and innovation. The European market is more fragmented and less competitive (prices do not fall after patent expiry) than the US one, European firms having to rely more on their domestic market to sell their products than on the whole European one.

Parallel to this, a report to the European Commission by the High Level Expert Group on the Intangible Economy (HLEG, 2000) argues that a key element of competitivity has become the exploitation of intangible investments such as R&D, proprietary know-how, employees' skills, world networks and brands and especially the capacity to combine external and internal sources of knowledge. Buigues et al. (2000) stress that intangibles such as R&D, marketing, advertising software and training, are growing in importance and have transformed the sources of competitiveness, so much so that public policies should change. They claim that public policies should shift focus towards sustaining intangible investments rather than tangible ones, in particular sustaining R&D and training.

Therefore intangible assets are being identified by some scholars as sources of competitiveness and differentials in economic performance. Given that innovation is about the creation of knowledge and that the capacity to commercialize innovation is determined by intangible resources such as organization (coordination between researchers, between researchers and marketers, motivation and so on), we investigate in this chapter whether the source of the difference in competitiveness between European and American pharmaceutical firms lies in intangible assets.

We first analyse intangible assets, in order to provide new evidence on their nature and their effects on competitiveness. Intangible assets are not new to economics and have been considered in this discipline; what is new is the acknowledgment of their growing importance in productivity at firm level and in growth at macro level. The Health Industry Model (HIM) implies that the appropriate production model for the health industry is a multi-product function; we argue in this chapter that this multi-product function includes intangible assets and the point is that, given their importance, it is high time to develop proper measurement of such assets, especially collective intangible assets which are the source of the externalities in the system.

The chapter is structured as follows. In the next section, we analyse intangible assets from the point of view of economic theory and show that, without their proper measurement, growth and productivity are underestimated. We see intangibles as capabilities resulting from complementary investments in both tangible and intangible resources. For instance, the development of human capital, innovation and the organization of production are complementary strategies aiming at enhancing the firm's major capability (producing the right product at the right moment, which is the intangible asset of the firm). The third section shows that intangible assets are particularly important in high-tech industries such as the pharmaceutical one. We show that the difference in competitiveness between the USA and Europe can be put down to a difference in intangible assets and a lack of exploitation of complementarities between assets. Policy implications are discussed in the conclusion. One major implication of our chapter is provision of a rationale for the rise in spillovers between the various components of the health industry and the resulting need for an integrated view, as proposed by Di Tommaso and Schweitzer in Chapter 1.

2 THE NATURE AND EFFECTS OF INTANGIBLE ASSETS

The increasing interest in intangibles as factors for competitiveness in the 'new' or 'information' or 'knowledge' economy is now quite obvious (Buigues et al., 2000; Lev, 2000). Just as obvious is the fact that intangible assets are not new. What is new is the importance they have assumed in recent years. Most often quoted factors are, first, the steep rise in the market-to-book value of many firms parallel to the small rise in the value of their physical assets; the difference is claimed to result from intangible assets. Second, globalization (hence more intense competition) and the diffusion of information and communication technologies favour the rising importance of intangible assets in that the former implies the increasing use by firms of non-price strategies (differentiation, product innovation), while the latter favours directly the development of intangible activities (more highly skilled personnel and services). A third factor is the diffusion of new organizational practices. Firms are delayering, redefining jobs towards multi-tasking and teamwork, and more interactions among employees. Communication within and between firms is increasing (in the latter case because of the process of outsourcing and collaboration with other firms, including competitors). This raises the issue of whether the diffusion of new organizational forms is the result or the cause of the rise in intangible assets.

Intangibles in Economics

Intangible assets have been considered in economics, but theories and measurement incur a number of problems. As long as production is mainly based on tangible assets (quantity of labour and capital) this is not a problem, but when such assets are key determinants of productivity and growth, the lack of proper measurement is a problem.

It can be argued that in general in the economy intangible assets have risen in importance. Competition has become tougher in many industries, for various reasons. First, globalization and the diffusion of information technologies imply that a firm can be threatened even locally by new entrants and the need to use non-price strategies. Second, the long-term evolution of some industries has led them to reach maturity or post-maturity phases in which both static and dynamic economies of scope can be achieved. Thus product renewal and quality have become key competitive factors in many industries. Innovation, creativity (design), organization and human capital are the major assets that generate quality and product renewal. Along the production process, this means that the manufacturing phase assumes minor importance relative to the research and the marketing phases; in the latter phases, the intangible assets mentioned above are essential.

Economics has considered intangible assets, such as innovation, knowledge, human capital and organization, but all rather separately. Traditionally, innovation studies have focused mainly on R&D activities and on process innovation, with the expenditure on R&D assumed to determine the rate of innovation, and on the determination of optimal patent (see Malerba, 2000, ch. 14, for a review; also Gilbert and Shapiro, 1990). The basis of the economics of innovation is the analysis of the market failures associated with the market for ideas. Geroski (1995) outlines three main market failures associated with the market for ideas. The first is spillovers (externality) arising because of the public-good character of knowledge. Thus knowledge is in many respects both non-rival (the use by one agent does not impede the use by another agent) and non-excludable (the producer of new knowledge cannot prevent other agents using the knowledge, although they do not pay for it). This creates the problem of appropriability, in that it is difficult to prevent other agents from taking advantage of a given innovation. The evolutionary theory disagrees with the non-excludability of knowledge, arguing that some knowledge (tacit) is difficult to imitate (Patel and Pavitt, 1995). The second market failure is non-convexity due to increasing returns (innovation has large fixed costs but low or zero marginal costs): marginal cost is then lower than average costs, and marginal cost pricing is not viable (firms therefore try to monopolize the market). Third, risk and uncertainty are inherent in the innovation process, which may lead to underinvestment in innovation. There are three types of uncertainties in R&D: technological (is it possible to do what we would like to do?), commercial (will there be a market for this new product?) and competitive (will competitors produce better innovation?).

A resulting problem is that it is difficult to price ideas: it is difficult to give a price to an idea before knowing it. However, once one knows the idea, there is no need to buy it any longer (Arrow, 1974). Hence the market tends to undervalue innovation, which leads to low returns to innovation and therefore reduces the incentives to innovate. Arrow (1962) made the first steps in formalizing the economic incentives to innovate within an equilibrium framework. He showed that incentives to innovate are higher under competition than under monopoly.

The public policy problem is that it is efficient to have maximum diffusion of knowledge, since marginal costs are nil, but maximum diffusion means low incentives to innovate (trade-off). The problem of spillovers has been at the heart of policy recommendations by economists: suggested policies have been subsidies (in order to maintain incentives to innovate and ensure diffusion), R&D cooperation and patents. Patents provide legal protection of the innovation. However, they have been shown in the empirical literature (Griliches, 1990, Patel and Pavitt, 1995) to be undervalued and imperfect since, in particular, they do not protect process innovation, the latter being better protected by lead time, secrecy and first-mover advantages.

In formal models (see Beath et al., 1995; Katsoulacos and Ulph, 1996; De Bondt, 1996; for reviews), spillovers are assumed to be a parameter that allow innovation by one firm to have effects (via cost reduction) on other firms. The parameter is generally exogenously given and there is no clear story on its determinants. Empirical research has attempted to measure spillovers but without much success. Generally R&D by other firms in the industry is added to the productivity equation (that measures the productivity effects of R&D). If the variable turns out to have a significant coefficient, there is spillover (defined as knowledge flow not accounted for in transactions; that is, externalities). Such a procedure is imperfect in that it incurs problems of aggregation (the spillovers may arise between two firms only and not between all firms in the industry) and it excludes some spillovers, such as the spillovers arising between industries and between the firm and its suppliers. In addition some knowledge flow can be unintentional and not external and therefore there is no problem for the return to innovation. In fact, in order to measure spillovers, one would need to evaluate the importance of different channels through which information flows: publications, employees changing firm, reverse engineering, and so on. The extent to which knowledge flows through these channels depends on the capacity of the receiver, the nature of knowledge and the incentives of individuals. Levin et al. (1987) have provided empirical evidence on the efficacy of a number of knowledge flows in their survey of the conditions for appropriability. They found that independent R&D is often cited as an effective way to learn knowledge from rivals, as well as licences. Other studies show that such channels are numerous and varied and differ according to the sector and the dimension of the firms.

Regarding organization, its consideration as a factor in differential performance across firms has recently experienced renewed interest, owing to increasing evidence of diffusing organizational changes in firms of most countries. The evidence results from surveys asking firms' managers about various aspects of organizational and technological changes. For instance, the COI survey in France was conducted on a sample of 5000 firms (Greenan and Mairesse, 1999). Other evidence includes Black and Lynch, 1997; Ichniowski et al., 1997; Ichniowski and Shaw, 1999, for the USA (although these surveys are more focused on human resources management systems, that is, internal labour markets).

Firms are becoming more decentralized, hierarchies are flattening, employees are involved in teams, rotate jobs and have increasing responsibility in problem solving that used to be performed by superiors. Communication within and between firms is increasing. Hence the traditional model of the multidivisional firm, a large, integrated and centralized firm, is no longer the dominant model, and this has attracted increasing interest among economists. However, the theory of the firm is made of different approaches which do not provide a complete framework to study organizational changes, in that no theory predicts why organizational changes occur, why they comprise different organizational practices (for instance, teamwork, together with greater communication, a flatter hierarchy and job rotation – not teamwork with less communication and fixed job positions). There is growing evidence that organizational changes are related to innovation and technological changes (Labory, 2000).

One interesting theory is that of complementarities, developed by Milgrom and Roberts (1995). According to them, firms adopt different but specific organizational practices together because those practices are complementary, in the sense of combining to produce positive profit effects. This is formalized in the supermodular properties of the profit function, the idea being that a practice is adopted jointly with other practices if this raises the marginal profit generated by the other practices. Complementarity is an interesting concept; as we will see below, it might be the missing piece of the puzzle necessary to analyse intangible assets.

In some empirical studies, the complementarity concept is claimed to provide the theoretical framework for organizational changes (Ichniowski and Shaw, 1999). All empirical studies using surveys of organizational changes analyse the effects of such changes on productivity by adding dummies for organizational practices to the production function, which is estimated (for example, Black and Lynch, 2000; Leoni et al., 2003). This procedure can be related to the multi-product function of the HIM. Not only do firms produce various outputs (products, services) but also they use various inputs, some of which have been neglected in the past, because they were not so important: intangible assets. The same studies of organizational changes generally also stress the importance of human capital, not in terms of quantity of labour but rather in terms of quality: the skill level of the workforce and the use of training programmes by the firm generally have positive effects on productivity (Black and Lynch, 1995; Bartel, 1992). In fact a rising level of human capital is associated with organizational changes: when the firm needs to increase quality and innovation, organizational changes are required to create new teams and more interactions in the firm, but also more skills are required (a workforce with higher levels of knowledge and competence).

In terms of measurement of value and effects, there are also imperfections. Innovation is a creation of ideas and is difficult to measure; as outlined above, a firm's innovative activity has been proxied in economics by R&D expenditure and patents. Organizational change is difficult to measure, yet it might be an important determinant of a firm's performance (as outlined by Leibenstein as long ago as 1966); many scholars have shown in case studies that the way the firm organizes production and the research team affects its performance in terms of efficiency of production and product innovation (for instance, there is the famous case of the Japanese car industry in the 1970s and 1980s; see Clark and Fujimoto, 1991; Nonaka, 1991; Labory, 1997). Human capital has generally been proxied by the education level, thereby excluding competencies gained during the working life through work experience and training within the firm.

In reality one might argue that all these intangible assets (innovation, organization and human capital) combine to create a firm's value: innovation (new knowledge) does not have value unless it is transformed into a product that is sold on the market. In other words, innovation, the organization of production, the organization of commercial activities and employees (with their human capital) all contribute to the value created by this innovation. The organization of the firm does not have value unless combined with human capital and tangible capital (such as machines and equipment). Hence it might be argued that the growing emphasis put by firms on their intangible assets (in audit, reporting and so on) is due to a strategy of development of complementarities between intangibles and tangible resources in order to increase value. Whereas in the past such complementarities were fixed (in particular product definition, technology, the organizational structure, job definition and skills of personnel), they tend now to constitute a strategic variable (Bianchi and Labory, 2002).

Intangibles for Firms

In the management literature, intangibles are defined as 'sources of probable future economic profits lacking physical substance, which are controlled, or at least influenced, by a firm as a result of previous events and transactions (self-production, purchase or any other type of acquisition) and may or may not be sold separately from other corporate assets' (Garcia-Ayuso, 2001, p.5). Interest has grown for intangible assets at firm level since many firms have attempted to develop methods to better report their intangible assets (see Zambon, 2002, for a review). There is a lack of a commonly shared measure but one concept that has emerged as useful to represent intangible assets in accounting is intellectual capital. Intellectual capital is argued to be composed of three components (Garcia-Ayuso, 2001):

- human capital, defined as the knowledge, skills, experience and abilities that employees have and that the firm does not own (and loses if the employee leaves the firm); examples are innovative capacity, creativity, know-how, professional experience, employee flexibility, motivation, satisfaction, learning capacity and loyalty;
- structural capital, that is, the pool of knowledge that stays in the firm when employees leave (organizational routines and procedures,

systems, culture and so on); examples are innovation capacity and organization flexibility;
- relational capital, which consists of the resources related to the external relationships of the firm, such as those with customers, suppliers and R&D partners; examples include image, alliances, customer loyalty, customer satisfaction, market power and environmental activities.

Intellectual capital and intangible capital are equivalent concepts. The set of intangible resources (stock, static notion) constitutes the intangible capital, while intangible investments (flow, dynamic notion) allow an increase in the intangible capital of a firm, through the acquisition of new intangible resources (such as new technology) or the improvement of the existing intangible capital (for example, organizational restructuring that improves communication flows within the company). Intangible investments may give rise to new intangible resources or may improve the value of existing ones.

Lev has proposed a measure of the long-term expected returns on knowledge assets. Knowledge capital is defined as the ratio between normalized earnings (several years of historical year-end results) minus earnings from tangible and financial assets and the knowledge capital discount rate. Table 6A.1 shows estimated values of the knowledge capital of major US pharmaceutical firms. However, this measure says nothing about the way intangible resources and investment create value. What is the role of each component of intangible capital, namely human resources, structural capital and relational capital? Part of the sociology literature (the 'fit' approach: Hannan and Freeman, 1977; Hannan, 1991) supports the view that all components create value only if they are jointly adopted (more than the sum of the parts). Such literature even refers to the Milgrom and Roberts (1995) complementarity concept.

The strategic management literature also agrees on the relationship between organizational and human resources management choices and the firm's main strategies. Thus, for instance, Arthur (1992, 1994) shows the correlation between industrial relations systems (more or less participative) and the market strategy of the firm (homogeneous good or not). Garcia-Ayuso (2001) argues that a firm's reporting on intangible assets starts with the definition of its main strategies, from which relevant intangible resources and investment (intellectual capital) result. Abernethy and Thomson (2001) try to find empirical evidence of the association between a strategy based on product innovation, organizational flexibility (the adoption of an organic structure, characterized by flat hierarchy, intense communication and decentralization of decisions), the importance of training and selection in human resource management and a lower use of management control systems. Contrary to expectations, they find that innovative firms do use management control systems, together with organic structure, training and selection. They also find evidence of an association between strategy and intangible resources and investments. However, their results have to be interpreted with care because they use a small sample, of only innovative firms with 200 employees and more.

A Firm's Value and its Intangible Determinants

Overall it seems that what has changed in recent decades and has led to the focus on firms' intangible assets is the firms' value creation process. The tangible capital appears no longer sufficient to explain a firm's value (see Table 6.1). It is therefore important to include such assets in the productivity (and growth) analyses, but the problem is that they are difficult to measure, given that often they are not priced and there is no market for them (organizational changes).

When firms renew products infrequently and rarely change strategy and organization, their value can be summarized by what they have achieved up to now: current market shares, products, investments and so on. Hence the traditional measures of performance, profits and revenues that come out of

Table 6.1 A taxonomy of the determinants of a firm's value

Tangible resources
Machines and equipment
Buildings
Labour (number of hours worked)
Existing products
Patents
Brand name
Distribution channels
Licensing agreements
Intangible resources
Human capital: competencies of employees
Internal organizational structure
Products in the pipeline
Relationships with other firms: joint ventures, alliances, collaborative agreements and so on
Relationships with public institutions: universities, local government and so on
Copyright, design rights, trade secrets
Know-how

the traditional balance sheet are good indicators. In contrast, when the firm renews products frequently and innovates regularly, current achievements do not summarize well the firm's value: what is in the pipeline is as important. This means that the future expected value should be taken into account when valuing a firm. However, such future value is by definition difficult to measure, and the best approximation is a probability distribution of future profits. In order to influence such probabilistic valuation, the firm has to build a reputation, a reputation for competence, so that market analysts and other stakeholders better perceive the intangible capital (the capabilities) of the firm.

If this is true, it also means that, at the country level, macroeconomic indicators and especially growth measures should account for these intangible assets. A country with high intangible assets has the potential to develop in the future, since its business has products in the pipeline and a capability to develop new products, create or conquer new markets. A country with low intangible capital has limited growth prospects.

Another element of the difficulty of measuring intangibles is that their effects are not only individual, but also collective: they are complementary or create complementarities (for instance, human capital is complementary to organization and knowledge in creating innovation). Hence an important asset is that constituted by the relationships (which in turn are often created by the organization). Such relationships are the source of the externalities of a system, and the health system (that is, the HIM) in particular. Relationships between individuals constitute the social capital of these individuals and/or their organization.

The next step is to analyse the pharmaceutical industry in the light of the above analysis; we will analyse the recent evolution of the industry, the relative performance of European firms and American firms in order to derive insights as to the differences in intangible assets and policy implications.

3 INTANGIBLE ASSETS IN THE PHARMACEUTICAL INDUSTRY

Since the beginnings of the industry in the late 19th century, pharmaceutical firms have followed various strategic orientations and built different innovative capabilities. Product innovation has always been the main competitive strategy, allowing a monopolistic position during patent life and generics competition afterwards. Market structure has been characterized by a few firms in oligopolistic competition together with a competitive fringe (competing in the generics market). A major change occurred with the development of biotechnologies, allowing new ways of doing R&D and new innovative capabilities. Major changes occurred in the organization of firms in the industry, although the main established firms maintained their position.

The Pharmaceutical Industry before Biotechnologies

Until the diffusion of the new biotechnologies, the dominant research technology was 'random screening' (whereby 'natural and chemically derived compounds are randomly screened in test tube experiments and laboratory animals for potential therapeutic activity': Henderson et al., 1999, p. 272). There were few knowledge spillovers between firms because what was important in the research was the quantity of research (screening) performed (a sort of economies of scale in research) and the basic scientific knowledge was shared by all competitors.

The industry developed especially after World War II, when large pharmaceutical firms consolidated their positions: for example, Merck, Eli Lilly, Bristol-Myers and Pfizer in the USA, Bayer in Germany, Hoffman-La Roche in Switzerland. R&D was institutionalized in-house, and large firms took advantage of economies of scale in research.

Public policy has always been substantial in the sector. Public research increased substantially after World War II. Patent protection has constituted another field of public intervention, as well as product regulation, with the obligation for all producers to submit new products for regulatory approval before being allowed to be commercialized. In the USA, the Food and Drug Administration (FDA) receives applications and decides on market authorizations; in the EU, market authorizations have only recently (in the 1990s) been harmonized, to allow producers to make just one application for market authorization in the EU, and not one authorization in each member state (see Bianchi and Labory, 2002). Another important difference between the USA and Europe concerns biomedical research. In continental Europe, the training and careers of scientists have been strongly oriented towards patient care and the application of research, because of the integration of medical schools and hospitals in single entities. In contrast, scientists in the USA and the UK have developed more expertise in basic research, since medical schools are separated from hospitals (Henderson et al., 1999). In addition, the links between universities (research institutions) and firms are stronger, in part thanks to a higher flexibility in the career of scientists, who can easily go to work for a firm for a certain time and come back to teach and do research at university afterwards if they wish. This has probably favoured the commercialization of innovations, a capacity which seems to be lacking in Europe (Gambardella et al., 2000).

Biotechnologies and Implied Changes in Industrial Organization

The fundamental discoveries in genetic engineering and recombinant DNA prompted the development of biotechnologies, which opened up new drug development opportunities for pharmaceutical companies. Whereas in the past drugs were derived from natural sources or synthesized through organic chemical methods, drugs can now be developed by genetic engineering, so that the compounds which were impossible to derive in the past, such as proteins, can now be produced artificially, allowing a 'rational' research method rather than random screening. Such discoveries have had important consequences for the drug industry (Henderson et al., 1999; Gambardella et al., 2000), in particular the arrival of new entrants, expert and specializing in biotechnologies, and the reorganization of major large established firms to provide a capacity to develop and produce biotech products.

The world pharmaceutical market was worth € 393 million (at ex-factory prices) in 2000 (Efpia, 2001). The North American market experienced the fastest growth and remained the world's largest market, with a 43 per cent share, compared to 22 per cent for Western Europe. The European pharmaceutical industry produces € 121.4 million worth of products and employs more than 500 000 employees. The main pharmaceutical producers are indicated in Table 6A.2.

In terms of R&D, Europe is losing competitiveness relative to the USA, although it was a world leader in terms of R&D and innovation until 1990. Between 1990 and 1999, R&D investment in Europe doubled, while in the USA it was multiplied by 3.5. In 1990, the major Europe-based companies spent 73 per cent of their R&D budget in Europe, while in 1999 they spent only 59 per cent of their R&D expenditure there; the difference went to the USA, implying that the USA has become a main world R&D centre. The loss in competitiveness concerns primarily biotechnologies. Figure 6.1 shows that, in 2000, Europe remained the main provider of new products relative to the USA, if one considers both chemical and biological entities; however, focusing on new biological entities, the USA is in the lead.

Table 6.2 shows that the USA is leading in the biopharmaceutical sector, R&D expenditure and turnover of biotech companies being much higher in the USA.

Table 6.3 shows the number of firms which are actively patenting at the European Patent Office. American firms are definitely more active, even more than indicated here, as the figures exclude patenting activity in the USA. Table 6.4 shows the difference in institutions that innovate in biotechnologies. Newly founded firms are much more numerous in the USA than in Europe, and represent an important source of innovation. Public research plays a more important role in countries like France

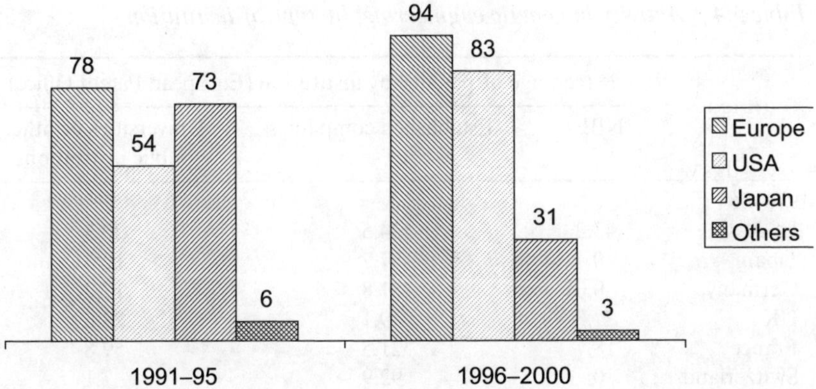

Source: EFPIA (2001).

Figure 6.1 New chemical or biological entities, 1991–2000

Table 6.2 Biotech companies: Europe versus USA

Indicators	Europe	USA
Turnover (Euro, m.)	8 679	23 750
R&D Expenditure (Euro, m.)	4 977	11 400
Number of companies (units)	1 570	1 273
Number of employees (units)	61 104	162 000

Source: EFPIA (2001).

Table 6.3 Patent applications at the European Patent Office

	World patent share (%) 1978–93	New firms (1978–86)	New firms (1987–93)	Growth in number of firms
USA	36.5	213	303	+90
Japan	19.5	108	185	+78
UK	5.9	39	64	+25
Germany	12.0	45	58	+13
France	6.0	37	52	+15
Switzerland	4.2	11	19	+8

Source: Henderson et al. (1999).

Table 6.4 Activity in genetic engineering, by type of institution

	Percentage of patents, by institution (European Patent Office)		
	NBFs	Established companies	Universities or other public institutions
1978–86			
USA	43.2	34.5	22.3
Japan	0	87.7	12.3
Germany	0.01	81.8	17.7
UK	27.3	49.1	23.6
France	18.7	21.5	59.8
Switzerland	0	92.9	7.1
Italy	0	95.7	4.3
1987–93			
USA	40.4	38.1	20.7
Japan	3.1	86.9	10.0
Germany	3.0	80.0	17.0
UK	23.7	44.7	31.6
France	16.7	35.0	48.3
Switzerland	4.7	89.0	6.3

Source: Henderson et al. (1999).

(see Chapter 12 of this book). Between the two sub-periods, the role of established firms has increased everywhere; this shows that the biotechnological sector initially grew with the arrival of new biotech firms (especially) in the USA, and progressively large established firms have developed a capacity to innovate in biotechnology. Henderson et al. (1999) show that large firms have acquired innovative capacity in biotechnology mainly by signing collaborative agreements, establishing research joint ventures or acquiring new biotech firms. Europe is progressively developing a capacity to innovate in biotechnology, but is a follower relative to the USA. In Europe, biotechnology has developed in different ways. Some firms, mainly the British and Swiss ones, have acquired biotechnologies via acquisition or collaborative arrangements with the smaller US biotech firms. Firms in other countries have benefited from public research, which has progressively caught up; thus, for instance, one of the biggest innovators in the EU is a French public institution, the Institut Pasteur.

Following the biotechnology 'revolution', R&D costs have continuously increased in the industry (see Table 6.5). R&D projects for new drugs nowadays last 12 to 14 years and cost up to $600 million (Bottazzi et al., 2000).

Table 6.5 *The changing structure of company costs in the pharmaceutical industry, 1973–89 (% of sales)*

	Manufacturing	Marketing	R&D	Operating profit	Other
1973	40	17	10	23	12
1973–80	37	16	11	27	11
1989	25	25	15	29	10

Source: Jacobzone (2000).

One interpretation of the collaborative agreements between established firms and new biotech firms is that large firms therefore have compensated for their lack of expertise in the new biotechnologies by developing agreements with the new entrants and by focusing on capabilities on the marketing and distribution side. Marketing costs have increased, and distribution strategies have become important determinants of market shares. Therefore complementary relations have developed between firms: new entrants with expertise on biotechnologies gain access to the distribution potential of the large firms, while large firms gain access to the innovative advantage of the smaller firms. Such complementary relationships generate collective intangible assets: the social capital that provides the basis for the development of other intangible assets, primarily innovation.

The USA has become a world R&D centre in biotechnologies; the percentage of new patents by new biotech firms (NBFs) has been the highest among OECD countries, with 43.2 per cent over the period 1978–86 and 40.4 per cent over the period 1987–93, while the respective figures were 34.5 per cent and 38.1 per cent for large, established companies. Over the two sub-periods, the percentage of new patents by large firms has increased, while that by NBFs has decreased. Parallel to this trend, the number of acquisitions of NBFs by large firms, or mergers between the two, has increased over the whole period. As a result, it appears that one way of acquiring biotech capabilities might have been, for large American pharmaceutical firms, the acquisition of innovative NBFs. As Table 6.4 shows, Germany and Japan have continued to rely primarily on large firms for innovation in genetic engineering (measured by patents), although some NBFs have started to innovate in the second sub-period. In the UK, the trend is different since universities and public institutions have increased their contribution to patents in genetic engineering over the period, while both NBFs and large firms have reduced theirs. The French pattern appears to be similar to the US one.

An account of the major mergers and acquisitions over the period 1987 to 1997 can be found in Bottazzi et al. (2000). Since then, further developments have taken place, with the creation of Novartis in Switzerland, and the merger of Hoechst and Rhône Poulenc. Mergers and acquisitions allow the pooling of resources and reduce the risks and costs associated with activities like R&D and the launch of new products. Given the large technological and organizational changes made necessary by the development of the biotechnologies, mergers and acquisitions among large pharmaceutical firms might be interpreted as a strategy to reduce the costs associated with the development of new complementarities and to control the new intangible assets.

This raises a policy dilemma: mergers and acquisitions may allow the acquisition of some capabilities in R&D which would be difficult to acquire otherwise, but they also raise the market power of firms; in fact, large incumbent firms threatened by the dynamism of NBFs in terms of innovation might have absorbed them also with a view to maintaining their market power. A deeper examination of such a dilemma would be useful, with an empirical study of mergers and acquisitions in the sector, for instance.

Apart from R&D, marketing and distribution costs play an important role in the costs of production. Distribution costs account for about 50 per cent of the expenditure on some products, when retail and wholesale margins are included, and vary by up to 10 per cent across member states (Huttin, 1989). Marketing costs are often higher than R&D costs (Jacobzone, 2000). The increase in R&D costs in the 1980s has been lower than the increase in marketing costs over the same period.

Intangible Assets in Europe and the USA

Hence we can provide a new key for interpretation of the evolution of the pharmaceutical industry. Following the significant technological change, new firms have entered the market thanks to their mastering of the new research capabilities. However, established firms, although constrained in their ability to adapt to changes, have maintained their dominant positions. In fact, established firms have adopted the new technology and therefore developed a capability to innovate and commercialize biotech products (intangible assets) by either setting up collaborative relationships with or acquiring the new firms, as in the USA, the UK and Switzerland, or taking advantage of the development of public research, as in France. The former strategy seems to have been the most profitable since the US firms have taken the lead in the introduction of new biotech products. The USA has become the world R&D centre, since many European firms have established new R&D centres in that country. The relevant tangible and intangible assets have converged there and complementarities have been built.

Hence the importance of proximity in creating complementarities and building intangible assets (as also stressed in Chapter 8 of this book).

The USA therefore has had major capabilities to adopt the new technology, to develop and commercialize new biotech products. Although in Europe biotech research is developing, it is claimed that what is lacking is the capability to commercialize innovation (Gambardella et al., 2000). Our analysis leads to the conclusion that, more importantly, what is lacking is the building of complementarities that allow the creation of intangible assets.

One way of favouring the creation of complementarities is to favour innovative networks. This has been an important measure taken to favour innovation and knowledge diffusion in Europe (Bianchi et al., 2002). Some networks have been created, but not always successfully. We show in another paper (Bianchi and Labory, 2003), that one important factor for the success of innovative clusters is the building of social capital, which favours relationships and exchange of ideas in particular. In terms of policy, this means providing the conditions for the social capital to develop. One such condition is to make sure that the bargaining power in the network be quite evenly distributed so that all actors can have access to the gains from innovation. One example of success in this respect is the BioRegion in Heidelberg (Germany).

Of course such an interpretation needs to be confirmed by a more systematic analysis; other interpretations include that of Thomas (1994), who argues that European firms adopted biotechnologies later because of weaker competitive pressures in their domestic market.

4 CONCLUDING REMARKS: COMPLEMENTARITIES IN THE HEALTH INDUSTRY

We have shown in this chapter that intangible assets might be defined as sets of complementary capabilities that result from various combinations of both tangible and intangible resources. Such assets appear to have been key assets for firms to adapt successfully to the new competitive conditions of the 1990s (mainly), with biotechnologies spurring the creation of new biotech start-ups active in drug development. Established large firms have managed to remain dominant players because of their control of a particular intangible asset: commercialization and distribution capabilities, which the new start-ups did not have. Among established firms, those who have taken a lead are those who early adopted the strategy to control the new technology, which they achieved by means of collaborative agreements with or acquisitions of new start-ups.

Our analysis has important policy implications, in particular concerning the creation of SMEs and the favouring of innovation and its diffusion. The chapter shows that what mattered for the former was not so much tangible resources, such as venture capital funding, but intangible ones, such as norms and institutional features (the social capital) favouring the development of intangibles: the different focus of biomedical schools not favouring applied research at the expense of basic research; the flexibility in scientists' careers and their possibility to move from university to business and back; and property right protection favouring innovation in universities. The absence or minor importance of such features in Europe appears to have led both to the delays in adopting the new technology and to the lack of commercialization of innovation. European governments and the European Commission are already stressing the need to provide a climate favourable to innovation and to firm creation for the respective policies to have positive effects. The consideration of intangible assets in policy making might lead to the formulation of policies effective in creating such a climate.

Regarding the latter point, policy making should look at the potential for complementarities to develop. As shown in this chapter, what matters with intangible assets is not so much the individual assets but the complementarities that arise between them. Thus the USA has become a world biotech R&D centre thanks to a combination of particular assets: university research oriented towards applied research; ease for researchers to set up firms and come back to university after working in the private sector; extensive business–university relationships, and so on. In Europe, the combination of assets is different; for instance, it is not so easy for researchers to start-up a business, and the training and research orientation do not always favour applied research. In addition, one has to consider the whole health industry in order to understand all the relevant complementarities: for instance, in France, the motivation of researchers and the orientation of universities can be understood only if the relationship between medical universities, that train personnel for medical research, the pharmaceutical sector and patient care in hospital, are considered. In the French system, medical faculties are strongly related to hospitals. In fact, each of them is associated with a particular hospital. This has the advantage of allowing doctors to train 'on-the-job'; however, the drawback is that medicine faculties are strongly oriented towards patient care and not so much towards research.

More generally, the importance of complementarities outlined in this chapter implies that restricting attention to a particular sector of health, say the pharmaceutical industry, or hospitals, will lead to the ignoring of important complementarities between the various sectors of the health industry. Thinking about the integrated health system allows us to see the accumulation of intangible assets in the whole system, not as externalities but as internalities. Considering the whole system, policy will make sure that all actors in the system have access to knowledge, innovation and human capital, which are internal to the system but external to the single actors. In particular, small firms also have access to information and knowledge in an integrated system, not in a disintegrated one.

Thus policies in different fields (education, research, industry and so on) have interactive effects that should be taken into account, because they might be negative for certain objectives. For instance, a (social) policy of reducing the prices of drugs combined with strong patent protection of new medicines might not be compatible. Another example is the case of the lack of transformation of innovation into commercial success in the EU; the policies to combine because jointly they have positive effects by creating appropriate intangible assets, are education (teaching at medical universities), RTD (Research and Technology Development which favours the links between universities and business, by allowing more mobility of researchers between the private and the public sector) and intellectual property rights (which allow university researchers to take out patents, so that they may have incentives to create new firms exploiting their innovation). The independent consideration of the policies does not resolve the problems: for instance, the ageing of the population would lead governments to take measures to increase the number of doctors, at the expense of researchers, while research and new and more efficient cures are also a solution.

Therefore what is needed is to create the intangible assets relevant and appropriate to the situation of the country (or supra- or infranational region), and for this purpose the integrated vision of the Health Industry Model is crucial. Hence the explicit consideration of intangible assets leads to strong support for the Health Industry Model; the externalities that the integrated view provided by the HIM points out are the complementarities outlined in this chapter.

A last note is on future research. More research is needed on the complementarities between assets, especially as regards the health industry, the social capital (including norms and institutions, hence laws, together with ethics) and other tangible and intangible capital such as knowledge and innovation. Thus, for instance, from 1980 onwards the Bayh–Dole Act gave universities the right to retain property rights from federally funded research, so allowing university researchers to enjoy the returns from their innovations. This probably favoured the creation of NBFs, as mentioned earlier. The Bush administration has changed this regulation, indirectly allowing the concentration of property rights in the hands of a few actors, with dramatic implications for innovation performance. This case is being investigated by the authors.

REFERENCES

Abernethy, M. and Thomson, G. (2001), 'Facilitating and managing knowledge creation in innovative firms: the role of management control systems', paper presented at the International Seminar on Intangibles and Intellectual Capital, Ferrara, June 2001.
Arrow, K. (1962), 'The economic implications of learning by doing', *Review of Economic Studies*, **29**, 155–73.
Arrow, K. (1974), *The Limits of Organisations*, New York: Norton.
Arthur, J. (1992), 'The link between business strategy and industrial relations systems in American steel minimills', *Industrial and Labour Relations Review*, **45**(3), 488–506.
Arthur, J. (1994), 'Effects of human resources systems on manufacturing performance and turnover', *Academy of Management Journal*, **37**, 670–87.
Bianchi, P. and Labory, S. (2002), 'The economics of intangibles', University of Ferrara, working paper no. 16/2002.
Bianchi, P. and Labory, S. (2003), 'Intangible assets nell'Industria Europea della Salute', *Rivista Italiana di Economia, Demografia e Statistica*, **LVI**(1–2), 281–98.
Bianchi, P., Iorio, R., Labory, S. and Malagoli, N. (2002), 'EU policies for innovation and knowledge diffusion', University of Ferrara, working paper no. 17/2002.
Bartel, A. (1992), 'Productivity gains from the implementation of employee training programs', NBER working paper no. 3893.
Beath, J., Katsoulacos, Y. and Ulph, D. (1995), 'Game-theoretic approaches to the modelling of technological change', in Stoneman, P. (ed.), *Handbook of the Economics of Innovation and Technological Change*, Oxford: Blackwell.
Black, S. and Lynch, L. (1995), 'Beyond the incidence of training: evidence from a national employers survey', NBER working paper no. 5231, August.
Black, S. and Lynch, L. (1997), 'How to compete: the impact of workplace practices and information technology on productivity', NBER working paper no. 6120.
Black, S. and Lynch, L. (2000), 'What's driving the new economy: the benefits of workplace innovation', NBER working paper no. 7479.
Bottazzi, G., Dosi, G., Lippi, M., Pammolli, F. and Riccaboni, M. (2000), 'Processes of corporate growth of an innovation-driven industry. The case of pharmaceuticals', European TSER Project.
Buigues, P., Jacquemin, A. and Marchipont J-F. (eds) (2000), *Competitiveness and the Value of Intangible Assets*, Cheltenham, UK and Northampton, MA, USA: Edward Elgar.
Clark, K. and Fujimoto, T. (1991), *Product Development Performance: Strategy, Organisation and Management in the World Automobile Industry*, Cambridge, MA: Harvard Business School Press.
De Bondt, R. (1996), 'Spillovers and innovative activities', *International Journal of Industrial Organisation*, **15**, 1–28.
EFPIA (European Federation of Pharmaceutical Industries and Associations) (2001), *Annual Report*, Bruxelles, www.efpia.org.
Gambardella, A., Orsenigo, L. and Pammolli, F. (2000), 'Global competitiveness in pharmaceuticals. A European perspective', report prepared for the Directorate General Enterprise of the European Commission, Bruxelles.
Garcia-Ayuso, M. (2001), 'Guidelines for managing and reporting on intangibles', Meritum's Project, TSER Programme.

Geroski, P. (1995), 'Markets for technology – knowledge, innovation and appropriability', in Stoneman, P. (ed.), *Handbook of the Economics of Innovation and Technological Change*, Oxford: Blackwell.

Gilbert, R. and Shapiro, C. (1990), 'Optimal patent length and breadth', *Rand Journal of Economics*, **21**, 106–12.

Greenan, N. and Mairesse, J. (1999), 'Organisational change in French manufacturing: what do we learn from firm representatives and from their employees?', NBER working paper no. 7287, August.

Griliches, Z. (1990), 'Patent statistics as economic indicators: a survey', *Journal of Economic Literature*, **XXVIII**, 1661–707.

Hannan, M.T. (1991), 'Theoretical and methodological issues in analysis of density-dependent legitimation in organisational evolution', *Sociological Methodology*, **21**, 1–42.

Hannan, M.T. and Freeman, J. (1977), 'The population ecology of organisations', *American Journal of Sociology*, **82**, 929–64.

Henderson, R., Orsenigo, L. and Pisano, G.P. (1999), 'The pharmaceutical industry and the revolution of molecular biology: exploring the interactions between scientific, institutional and organisational change', in Mowery, D.C. and Nelson, R. (eds), *The Sources of Industrial Leadership*, Cambridge, MA: Cambridge University Press.

HLEG (High Level Expert Group on the Intangible Economy) (2000), *The Intangible Economy – Impact and Policy Issues*, report of the European Commission's HLEG on the Intangible Economy, Luxembourg: CEC publications (March, ISBN 92-894-0019-6).

Huttin, C. (1989), 'The distribution of pharmaceuticals. An international survey', *Journal of Social and Administrative Pharmacy*, **6**(4).

Ichniowski, C. and Shaw, K. (1999), 'The effects of HRM systems on economic performance: an international comparison of US and Japan', *Management Science*, **45**(5), 704–21.

Ichniowski, C., Shaw, K. and Prennushi, G. (1997), 'The effects of HRM systems on productivity: a study of steel finishing lines', *American Economic Review*, **87**, 291–313.

Jacobzone, S. (2000), 'Pharmaceutical policies in OECD countries: reconciling social and industrial objectives', OECD Labour Market and Social Policy – Occasional Papers no. 40, Paris.

Katsoulacos, Y. and Ulph, D. (1996), 'Endogenous innovation spillovers and technology policy', Centre for Economic Policy Research, discussion paper 1407, May.

Labory, S. (1997), 'Firm structure and market structure in imperfectly competitive markets', PhD thesis, European University Institute, Florence.

Labory, S. (2000), 'Indirect effects of organisational innovations', paper presented at the EAEPE 2000 Conference, Berlin, 2–5 November.

Leibenstein, H. (1966), 'Allocative efficiency versus X-efficiency', *American Economic Review*, **56**, 392–415.

Leoni, R., Cristini, A., Gaj, A. and Labory, S. (2003), 'Flat hierarchical structures, bundles of new work practices and firm performance', *Rivista Italiana degli Economisti*, **2**, 131–41.

Lev, B. (2000), 'Intangibles. Management, measurement and reporting', Stern School of Business, New York University, December.

Levin, R., Klevorick, A., Nelson, R. and Winter, S. (1987), 'Appropriating the returns from industrial research and development', *Brooking Papers on Economic Activity*, **3**, 783–820.

Malerba, F. (2000), *The Economics of Innovation*, Rome: Carocci.
Milgrom, P. and Roberts, J. (1995), 'Complementarities and firms: strategy, structure and organisational change in manufacturing', *Journal of Accounting and Economics*, **19**, 179–208.
Nonaka, I. (1991), 'The knowledge-creating company', *Harvard Business Review*, November–December, 98–104.
Patel, P. and Pavitt, K. (1995), 'Patterns of technological change', in Stoneman, P. (ed.), *Handbook of the Economics of Innovation and Technological Change*, Oxford: Blackwell.
Thomas, L. (1994), 'Implicit industrial policy: the triumph of britain and the failure of France in global pharmaceuticals', *Industrial and Corporate Change*, **3**(2), 451–89.
Zambon, S. (2002), 'Accounting, intangibles and intellectual capital: an overview of the issues and some considerations', PRISM Project, European Commission, DG Information Society (http://www.Euintangibles.net).

APPENDIX

Table 6A.1 Lev's estimation of knowledge capital (in $ millions)

Pharmaceutical companies	Knowledge capital	Book value	Knowledge capital/ book value	Market value	Mkt/ book value	Knowledge capital earnings/ sales (%)
Merck	48 038	12 614	3.81	139 910	11.09	22
Bristol-Myers Squibb	30 470	7 219	4.22	106 994	14.82	19
Johnson & Johnson	28 965	12 359	2.40	92 884	7.52	14
Pfizer	23 890	7 933	3.01	136 846	17.25	20
American Home Prod.	22 822	8 175	2.79	63 392	7.75	17
Abbott Labs	19 558	4 999	3.91	56 631	11.33	17
Eli Lilly	16 505	4 646	3.55	67 968	14.63	21
Warner Lambert	12 099	2 836	4.27	52 237	18.42	16
Pharmacia & Upjohn	4 725	5 538	0.85	22 447	4.05	7
ICN Pharmaceuticals	1 158	796	1.45	3 092	3.88	16
Watson Pharmaceuticals	1 110	565	1.96	3 899	6.90	35
Allergan	1 053	841	1.25	2 753	3.27	10
Mylan Labs	972	744	1.31	3 666	4.92	19
Rexall Sundown	766	192	4.00	2 392	12.48	31
Alza	622	301	2.07	4 181	13.88	14
Forest Labs	553	614	0.90	2 653	4.32	14
NBTY	386	117	3.30	976	8.34	15
Barr Labs	376	156	2.41	909	5.83	11
Perrigo	254	426	0.60	821	1.93	3
Agouron Pharmaceuticals	152	236	0.64	1 049	4.44	3

Table 6A.2 Top ranking companies in the health industry (ordered by amount of R&D spending)

	R&D per employee ($000)	Country of origin
Chemicals		
Bayer	12.6	Germany
Du Pont de Nemours	12.8	USA
BASF	9.2	Germany
Dow Chemical	14.3	USA
AKZO Nobel	7.3	Netherlands
Mitsubishi Chemical	11.7	Japan
Sumitomo Chemical	19.9	Japan
Toray Industries	6.4	Japan
Solvay	8.4	Belgium
Mitsui Chemicals	19.1	Japan
Asahi Chemical Industry	8.4	Japan
PPG Industries	5.7	USA
ICI	3.8	UK
Rohm and Haas	9.4	USA
Clariant	5.4	Switzerland
Teijin	7.6	Japan
DSM	7.7	Netherlands
Shin-Etsu Chemical	8.4	Japan
Kyowa Hakko Chemical	19.3	Japan
Dow Corning	13.2	USA
Rhodia	4.1	France
Ciba Speciality Chemical	6.0	Switzerland
Health		
Abott Laboratories	14.9	USA
Warner Lambert	19.2	USA
Medtronic	14.2	USA
Baxter International	5.9	USA
Guidant	25.6	USA
Applera	37.8	USA
Becton Dickinson	6.0	USA
Nycomed Amersham	17.0	UK
Boston Scientific	9.7	USA
Allergan	21.2	USA
Beckman Coulter	12.8	USA
Pharmaceuticals		
Pfizer	33.0	USA
GlaxoSmithKline	23.3	UK
Johnson & Johnson	19.9	USA

Table 6A.2 (continued)

	R&D per employee ($000)	Country of origin
AstraZeneca	34.0	UK
Novartis	23.2	Switzerland
Pharmacia	31.2	USA
Roche	25.2	Switzerland
Merck	22.6	USA
Eli Lilly	37.9	USA
Bristol-Myers Squibb	29.5	USA
American Home Products	23.5	USA
Aventis	9.8	France
Schering-Plough	31.8	USA
Boehringer Ingelheim	22.3	Germany
Sanofi-Synthelabo	20.3	France
Amgen	77.2	USA
Schering	21.8	Germany
Takeda Chemical	27.9	Japan
Sankyo	33.3	Japan
E Merck	10.2	Germany
Yamanouchi Pharmaceutical	35.9	Japan
Novo Nordisk	22.5	Denmark
Eisai	39.1	Japan
Fujisawa Pharmaceutical	33.4	Japan
Chugai Pharmaceutical	48.1	Japan
Daiichi Pharmaceutical	30.6	Japan
Elan	61.0	Ireland
Biogen	137.4	USA
Chiron	57.3	USA
Millennium Pharmaceutical	135.3	USA
Serono	41.3	Switzerland
Shionogi	n.a.	Japan
Altana	16.1	Germany
Taisho Pharmaceutical	27.2	Japan
Incyte Genomics	97.5	USA
Lundbeck	39.7	Denmark
Ono Pharmaceutical	n.a.	Japan
UCB	12.0	Belgium

Note: n.a. = not available.

Source: *Financial Times*.

7. Benchmarking hospital costs in the UK: increasing efficiency and driving innovation in a health care industry?*

Sue Llewellyn and Deryl Northcott

INTRODUCTION

A Health Industry Model (HIM) recognizes the broad contribution that health care can make to national economic development. Yet, wherever and whenever health care is publicly financed, a key government policy objective is achieving technological development and innovation within resource constraints. Governments have employed a variety of mechanisms to achieve this aim. In the UK, for example, since the 1980s, health care management policy has undergone several reforms, from general management to resource management and the internal market through to, latterly, the 'New Labour' government's modernization programme. Several authors have charted the history of these developments (see, for example, Bourn and Ezzamel, 1986a; Broadbent et al., 1991; Broadbent, 1992; Preston et al., 1992; Harrison and Pollitt, 1994; Hood, 1995; Jones and Dewing, 1997; Jones, 1999; Llewellyn, 1998; Keen, 1999; Klein, 1999). In conjunction with these broader financial management agenda, a succession of costing initiatives have changed the way in which cost information is compiled, reported and used for control purposes in UK hospitals. Costing was reconfigured to support general management through devolved budgets, resource management through clinical budgeting and the internal market through setting prices equal to cost for contracting and commissioning (see Bourn and Ezzamel, 1986b; Bates and Brignall, 1993; King et al., 1994; Ellwood, 1996a, 1996b, 2000; Jones, 1999).

However, until the advent of 'modernization' and relative performance evaluation in UK health care, these financial management reforms and their associated costing regimes have been internally focused cost control measures. The introduction of benchmarking changes this focus by providing external markers for cost control. The UK New Labour government's 'modern and dependable' policy for the National Health Service (NHS) depended upon 'The new performance framework [that] will encourage greater benchmarking of performance in different areas, and the publication of comparative information [that] will allow people to compare performance and share best practice' (DoH, 1998a, p. 6). Hence the benchmarking approach anticipates that hospitals will, within a context of cost constraint, share best practice in order to drive technological innovation in clinical practice. This chapter explores how far this aim has been realized.

To focus the study the benchmarking intent of the most recent substantive NHS costing initiative, the national reference costing exercise (NRCE) and its associated cost index (NRCI), henceforth termed 'the Index', is examined. This initiative imposes a requirement on all English NHS acute Hospitals[1] to report their costs, on a consistent basis, for a wide range of health care activities. In order to compute reference costs and the associated Index, health care activities are categorized within health care resource groups (HRGs), the costs of which are calculated retrospectively, based on actual costs incurred by hospitals (DoH, 1997; NCMO, 1997a, 1997b). HRGs are a variant on the diagnostic related groups (DRGs) developed in the USA for pricing health care services. The National Casemix Office[2] constituted HRGs to 'group together treatments that are clinically similar, consume similar quantities of resources and are likely to be similar in cost' (DoH, 1998b, p. 4). The published Index league table is based on HRG reference costs and ranks hospitals on the basis of their apparent cost efficiency to create a 'ladder of success'. Since 1998, this Index has been made available throughout the NHS; it is also open to public scrutiny.

The scope and detail of costed HRG data presented as reference costs have been expanding. It is intended that by 2004 the NRCE will provide comprehensive cost data across all non-primary health care activities within any mode of service delivery (for example, elective and emergency inpatients, day cases, outpatients, critical care, accident and emergency, community care). The NRCE will then constitute the largest cost information resource ever made available to support NHS benchmarking, cost management and decision making.

It was anticipated that these new, comparative health care costs would be used by purchasers of health care services[3] and by the NHS executive as 'a strong lever with which to tackle inefficiency and differential performance' (DoH, 1998b, p. 1) by placing pressure on 'inefficient' hospitals to reduce costs. Another intention of the NRCE was to provide NHS hospitals with 'the opportunity to identify cost differences and . . . understand the reasons behind them' (DoH, 1998c, p. 3), the underlying assumption being that cost variability reflects undesirable inefficiency in those hospitals where costs are higher. The NHS executive advised that 'individual NHS Trusts [hospitals] should consider their position on the Index and discuss with their peers reasons for differences and the scope this may offer for efficiency savings' (ibid., p. 20). The stated aim of the NRCE is to make cost information 'visible' so that hospital managers can 'tackle unacceptable variations in performance and raise overall standards across the NHS . . . by sharing information and comparing performance' (ibid., p. 3). The sharing of cost information on relative performance was then expected to aid cost control by incentivizing each hospital's managers and clinicians to work together to identify best practice and reduce variations in costs and efficiency. As can be discerned, the benchmarking themes of rigour, measurement against a referent other, information sharing and hence continuous improvement are all present in the government rhetoric.

A news article entitled '*List of shame* to feature NHS Trust costs' (*The Financial Times*, 12/6/98) referred to the (then) new reference cost Index as 'a league table of efficiency' which would lead to 'the controversial naming and shaming of under achievers', with the focus placed clearly on apparent cost variability. Perceptions within government have been similar. In the same *Financial Times* article, the health minister of the time was reported as noting that a 'hip replacement on the NHS costs £2000 in some areas but more than £8000 in others'. He went on to state: 'There are currently some unacceptable variations between hospitals in the cost of treatments . . . . These new measures [published Index results] will iron them out and get better value for money for patients' (*Financial Times*, 12/6/98).

By 2000, the Minister of State for Health was proudly noting that the NRCE had assembled 'more cost data than any other reporting system anywhere else in the world' (DoH, 2000, p. 3). Yet, four years into the initiative, the reliability and comparability of reference cost data and, hence, its usefulness for benchmarking remain in question. Reference cost schedules published to date reveal wide variations in unit costs for almost every HRG, but there is still little understanding of the significance of this variability in terms of assessing the impact of technological innovation on health care costs.

One problem in using the NRCE and the Index for driving technological development in health care is that the NRCE represents a particular type of benchmarking. The Index positions multiple hospitals on a 'ladder of success' (Fitzpatrick and Huczynski, 1990; Mann et al., 1998). This type of benchmarking is external, comparative and results-oriented, external, as opposed to involving an internal comparison between similar units within an organization, comparative in the sense of making comparisons between direct competitors in the same sector and results-oriented (how other peoples' outputs compare with ours) rather than being process-oriented (how other people do things as compared to us).

This type of relative performance evaluation is gaining in popularity in the UK public sector (Bowerman et al., 2001; Northcott and Llewellyn, forthcoming). Yet, as Holloway et al. (1999, p. 2) point out, this form of benchmarking will only lead to performance enhancement (as opposed to efficiency gains only) if the external results orientation leads to an internal examination of processes, as it is only through a better understanding of how inputs are transformed into outputs that superior results can be obtained. Currently in UK health care, detailed information on hospital processes is not publicly available and, hence, comparative process exercises (with a view to technological development and innovation) are not possible on the basis of the 'ladder of success'. However a substantial proportion of UK hospitals[4] purchase the services of private sector consultancies to assist them in 'drilling down' from the mandatory benchmarking information to locate specialized benchmark activity reports that are used within comparative hospital cluster groups to explore issues of best practice.

Several and varying motivations may trigger benchmarking exercises; in a private sector context Wolfram Cox et al. (1997) suggest the following possible motivators: improvement, consolidation, inclusion/exclusion, compliance, learning/teaching/mentoring and financial gain. In the more regulated and political environment of the not-for-profit public sector, improvement, consolidation, compliance and learning/teaching/mentoring seem to be the more relevant factors. Definitions of benchmarking return to the recurrent themes of continuous improvement, measurement against a referent other, and rigour (ibid.). Camp (1989) adds the concept of 'information sharing' to the central motifs of benchmarking and, hence, points to the possibility of process enhancement through comparative exercises. It has been argued that, in the public sector, benchmarking is both more necessary because of the traditional lack of a results-oriented culture and, also, more difficult as staff have become accustomed to accepting rather than questioning the system (Bendell et al., 1993). Along with moves for greater accountability, the impetus from government towards greater transparency in services is central to the introduction of benchmarking in the public sector (for evidence of this intent in the health care sector, see Department of Health [DoH], 1998a, 1999a).

The structure of this chapter is as follows. The next section outlines the research method. The results of the published Index are then discussed. Potentialities and problems with this Index are explored and the conclusions reflect on the extent to which the Index is a useful tool for benchmarking, cost control and technological innovation. Finally, some possible further policy developments to enhance the role of benchmarking in health care are suggested.

RESEARCH METHOD

The research design for this study was multidimensional. Documentary material related to the Index and published Index results were obtained from the NHS executive. Case studies were carried out in six NHS hospital sites to explore perceptions of the Index and the issues involved in compiling reference cost data at hospital level. These hospitals were a mixture of large teaching, non-teaching metropolitan and non-metropolitan serving more disparate, rural populations. The six sites were spread around England: one in each of the North West, Trent, South West and London regions, and two in the South East.

Access was gained to documents and working papers produced within the costing divisions of these six studied hospitals and a range of personnel were interviewed within each: management accountants involved in compiling reference costs, the finance director, two clinical directors (from one surgical and one medical specialty area) and information management personnel involved in producing patient activity data. The management accountants and finance directors were interviewed with the dual aims of determining the processes they followed in compiling reference cost data, and of gathering their views on the Index and its associated reference costs. Clinical directors in both surgical and medical areas contributed their views on the nature of differing clinical procedures and their corresponding resource consumption and cost patterns, as well as commenting on their experiences and uses of cost data. The findings from these six case studies were combined with evidence from interviews with personnel at the NHS Executive Financial Management Branch involved in setting up the NRCE and with preparing relevant materials and directives, finance and commissioning personnel in selected regional health authorities, a senior financial analyst in a large NHS executive regional office and members of a private sector health care benchmarking agency.

The main tranche of interviews was carried out at the selected hospital sites (and in other locations in the cases of non-hospital interviewees) between December 1999 and September 2000. Further follow-up interviews were undertaken between November 2001 and February 2002. All interviews were semi-structured in nature and ranged from one to two hours in duration. They were tape-recorded and later transcribed.

In addition to the interview evidence, a survey was undertaken of all NHS hospitals included in the Index. This survey had a 42 per cent response rate and was carried out in October 2000. The aim of the survey was to gather a more generalized view of the usefulness of the Index than had been obtained from the interviews. The quotations used in the later sections of the chapter make reference to both interview and survey data.

THE NATIONAL SCHEDULE OF REFERENCE COSTS (NSRC) AND THE NATIONAL REFERENCE COST INDEX (NRCI)

Reference cost data are made available to NHS hospitals (and other users) in three formats considered useful for benchmarking and cost control. These are the National Reference Cost Index (NRCI), the National Schedule of Reference Costs (NSRC) and individual hospital HRG costs.

The NRCI (the Index) and the NSRC are the information sets referred to in this chapter and are the outputs most focused on in the media, by politicians and by NHS managers. Through giving visible quantification to the range of NHS cost variability, the Index is designed to facilitate hospital benchmarking (DoH, 1998b, p. 8). Nineteen hospital 'clusters' were identified (ibid., pp. 11–20) in the original reference cost documentation so that hospitals operating in similar environments, with a similar casemix, could benchmark against each other. However, these 'clusters' did not become operational in the published indices that followed. As discussed above, some hospitals have, however, located their own comparator institutions for benchmarking purposes and the identification of best practice, frequently assisted by the private sector benchmarking agency mentioned earlier.

The NRCI constitutes a benchmarked 'ladder of success' by presenting a single figure for each NHS hospital that 'compares the actual cost for its casemix with the same casemix calculated using national average costs' (DoH, 1998c, p. 17). An index score of 100 is interpreted as 'average' cost performance, whereas scores above or below 100 suggest above or below average cost performance respectively; for example, a score of 102 indicates costs that are 2 per cent above the average whereas a score of 98 may indicate a somewhat more efficient hospital performance. The NHS executive claims that this index measure gives 'an indication of the overall technical efficiency of a Trust [hospital]' (DoH, 1998b, p. 8).

As Table 7.1 demonstrates, 'the ladder of success' is a slippery one! There have been some surprising year-on-year movements up and down the ladder. For example, North Hampshire shot up to first place in the 2000 results, an astonishing jump from the previous year when it was virtually at the bottom of the table, and the Royal National Hospital for Rheumatic Diseases, a presentable 28th on the 1999 ranking, dropped to 118th in 2000. Overall the rankings demonstrate extreme instability in reported performances. What do these jumps and falls mean? Two contrasting possibilities can be identified. Either the Index is a compelling benchmarking tool and hospitals focus so intently on improving their rating that they are able to achieve remarkable results in one year only, or the data quality is so poor that 'leaping' and 'falling' hospitals appear as mere artefacts of the data collection process.

Table 7.1 The 'ladder of success' for NHS hospitals

NHS trust (hospital)	Market forces factor adjusted trimmed index			
	2000 index		1999 index	
	Score	Ranking[1]	Score	Ranking[1]
The 2000 'top five' trusts				
North Hampshire Loddon Community	63	1st	150	228th=
Moorfields Eye Hospital	71	2nd	88	28th=
Weston Area Health	73	3rd	80	7th=
Nottingham City Hospital	74	4th	100	113th=
Countess of Chester Hospital	74	5th	75	3rd
2000 'average' trusts[2]				
Portsmouth Hospitals	99	150th =	97	85th=
Grantham & District Hospital	100	151st =	108	168th=
City Hospitals [Birmingham]	100	151st =	92	49th=
Salford Royal Hospitals	100	151st =	113	188th=
North Hampshire Hospitals	100	151st =	96	74th=
East Kent Acute	100	151st =	n./a.[3]	n./a.
Tameside & Glossop Community and Priority Services	101	156th =	106	152nd=
The 'bottom five' trusts				
Royal National Orthopaedic Hospital	134	117th	154	232nd
Royal National Hospital for Rheumatic Diseases	145	118th	91	28th=
Preston Acute Hospitals	148	119th	103	129th=
Great Ormond Street Hospital for Children	164	220th	117	202nd=
Birmingham Children's Hospital	174	221st	134	226th=

Notes:
1. The higher the ranking, the lower the trust's relative index score (and the greater its assumed cost efficiency). A total of 221 trusts was included in the 2000 National Reference Cost Index; 236 trusts were included in 1999.
2. In the 2000 index, only five NHS trusts achieved exactly the 'average' score of 100. Closest to this 'average' were 12 NHS trusts with an index score of 99 (those not shown above are Birmingham Heartlands & Solihull; Royal Marsden Hospital; Hereford Hospitals; City Hospitals, Sunderland; Rotherham General Hospitals; Morecambe Bay Hospitals; Kingston Hospital; Worcester Royal Infirmary; Kent & Sussex Weald; Brighton Health Care; Newham Healthcare) and seven NHS trusts with an index score of 101 (those not shown above are Trafford Healthcare; Bolton Hospitals; Nuffield Orthopaedic; Dartford & Gravesham; Mid Kent Healthcare; Wiltshire & Swindon Healthcare).
3. Not included in the index for that year.

Table 7.2 Key national reference cost statistics

NRCI results	1997/98	1998/99	1999/00	2000/01	2001/02
NRCI range*	−33% to +62%	−33% to +86%	−37% to +74%	−46% to +112%	−39% to +99%
% of trusts within 20% of the average (100 score)**	90	86	87	82	88
% of trusts within 10% of the average (100 score)	60	61	62	58	72

Notes:

* All index statistics presented here are based on the trimmed index adjusted for 'market forces' (that is, differential regional costs). This is the index selected for comment in the published reference cost documents.

** This +/− 20 per cent range is highlighted in NHS executive reference cost publications.

Source: Compiled from data in DoH (1998b; 1999c; 2001b).

Table 7.2 shows that the phenomenon of cost variability has bedevilled the NRCE since its inception and, although in the latest published information the range of variability has narrowed somewhat (see the proportion of hospitals within 10 per cent of the average), substantive variation still remains. How useful is this 'ladder of success' and its associated reference costs for benchmarking purposes in the UK NHS? The following sections discuss this issue.

BENCHMARKING POTENTIALITIES FOR THE NRCI AND NSRC

Within provider organizations, there is a recognized need for information sharing in the NHS to enhance the dissemination of best practice and to drive technological development. As noted above, the Index was intended to assist NHS hospitals in benchmarking their cost efficiency against other hospitals, managing their costs and encouraging managers and clinicians to work together in reviewing the efficiency of services provided. Findings from this study suggest that most hospital actors are concerned to make the most of the Index for benchmarking, cost management and the sharing of best practice. Interview evidence revealed a measure of guarded optimism about the potential future role of reference cost information. For example:

> I would like to use reference costs more to understand the differences between hospitals. I would like to think of them as true costs, done the best that you

can, and then try to understand why is it still so different between one Trust [hospital] and another. (Interview: cost accountant, a Trent NHS hospital)

As discussed earlier, many UK hospitals hire the services of a private sector consultancy to assist them in using the standard reference cost information internally to draw up more detailed reports. Such moves indicate that the UK hospitals do take their reference costs seriously as a basis for further inquiry: 'We take small areas of activity and home in on them. Like prosthetic costs, something like that, to try to make more sense of the data' (interview: private sector benchmarking agency, member 1). These internal reports also constitute 'early warning systems', allowing the hospitals to take remedial action in advance of official reports.

> We provide Trusts [hospitals] with an early warning system . . . so that they can understand where they are in advance of the publication date so that if there is anything to be put right or to be alerted about, then they [the hospital] are there before the Department [of Health]. (Interview: private sector benchmarking agency, member 2).

The indications are that these detailed reports on hospital processes are being used more as internal early warning systems rather than disseminated more widely to facilitate the exchange of information between hospitals in benchmarking cluster groups.

BENCHMARKING PROBLEMS WITH THE NRCI AND NSRC

Although, as the above section indicates, both purchasers and hospitals see potentialities in the reference cost initiative, there are at present considerable question marks over the credibility of the reference costs for benchmarking. Case study and survey evidence revealed two key issues that confound the use of the Index for relative performance evaluation and the sharing of information on best practice: first, the absence of an accepted benchmark for relative performance evaluation; second, problems relating to institutions that consider themselves exceptional. These issues are discussed, in turn, below.

Absence of Standards

The biggest problem with the Index for benchmarking purposes is the lack of a benchmark! In the absence of such a benchmark (or standard), hospitals aim for an average or 'normal' performance and the norm becomes the standard aimed for. 'Everyone wants to be at 100 on the index rather than being 120 or 80, so everyone is clustering around the norm, rather than trying to aim for a standard' (interview: clinical director, a South East NHS hospital). There is no indication of what an excellent performance would be. Is being the lowest 'scorer' on the Index a superlative outcome? Equally, is having the highest Index a matter for hand wringing, or is it indicative of being a high-quality, technologically advanced (and, hence, high-cost) centre of excellence?

> The people at the poles should be analysed in more detail. Those at the bottom [of the index] – is that because they are super-efficient, so there are lessons to be learnt and the average should come down, or is it that they are not doing things that they should? Equally the people at the top, costing heaps – is it because they are giving a Rolls Royce service, or are they just completely inefficient? (Interview: clinical director, a Trent NHS hospital)

In the absence of both a benchmark for excellence and a definition of unacceptable performance, hospitals are free to offer their own interpretations of their Index scores. One cost accountant noted with irony: 'A low index score means you're efficient, but a high index score means that your quality must be excellent!' (interview: cost accountant, a South West NHS hospital). Maybe the benchmark or standard does not equate to an excellent performance but a 'good enough' one: the hospitals are confused. 'If we are talking about standards, are we talking about minimal acceptable standards or are we talking about best practice? We haven't really sorted that out in our minds' (interview: clinical director, a Trent NHS hospital).

The government has encouraged hospitals to focus on cost comparisons with similar institutions. However, in the absence of any 'best practice' or 'good-enough' standards, any new information from comparator hospitals merely tends to shift perceptions of 'good' performance.

> Last year we came in at 98. And 98 is probably a fairly good place for us to be, as we are a very large acute Trust [hospital] just outside London. But our sister Trust down the road came in at 89 so suddenly we seemed to be very inefficient. (Interview, cost accountant, a South West NHS hospital)

The problem of lack of standards for reference costs (what is the referent?) is confounded by the NHS executive's recognition that cost efficiency is not, of itself, an indicator of good performance, but requires interpretation in the context of clinical quality indicators and knowledge of local circumstances. A finance director notes: 'There's no external standard. All one's got is this [cost] benchmark, and its biggest failing is that it's not clear what the acceptable clinical standard is, to measure it against' (interview: finance director, a South West NHS hospital). Without clear indicators of clinical performance, hospital actors find it difficult to interpret whether reference cost differentials indicate a need for cost control, or are justified on other grounds such as quality of care. The greatest problem with the 'reference' costs is the lack of a 'referent other' for benchmarking.

Issues relating to 'Exceptional' Cases

For benchmarking to be seen as legitimate, the benchmarkees have to be similar in the relevant respects. Arguments abound from participants in this study regarding the 'exceptional' nature of their circumstances and, hence, the inappropriateness of their inclusion in the Index. The most obvious instance of hospitals that consider themselves to operate in exceptional circumstances and, hence, to be difficult candidates for benchmarking is when the institution concerned is a specialist one.

> At X I think its [the Index] use is very limited as we are a specialist paediatric centre. Previously I was at a general London Trust [hospital] and the reference costs were a lot more useful to a big Trust like that. The normal paediatric HRG groupings are not much help to a specialist paediatric centre. Although I know that they have done work to go down one level, so there is now one speciality paediatrics [grouping]. But that's not enough when you are a specialist Trust. The use of benchmarking and comparing yourself is not the best tool for us because we are very high cost with everything sort of pushed into very few HRGs. (Interview: finance director, a London NHS hospital)

Such specialist hospitals are almost inevitably going to resist comparisons that they judge to be unfair: 'It isn't good for us when people look at a league table or look at our costs on an index compared to other hospitals that may have children's paediatrics' (interview: finance director, a London NHS hospital). This finance director prefers to choose her own benchmarking group that she feels is more appropriate – albeit that any comparator group is going to be problematic.

> We are in a benchmark group with other children's hospitals – Birmingham, Alderhey, Manchester and Sheffield – but they have not got the London factor. So I find that benchmarking ourselves with other London teaching hospitals is better because they have got the London high cost, staffing issues and the fact that we are so heavily reliant on agency nurses. So I have been trying to do some benchmarking with them and whilst they have only got a small element of paediatrics, that is probably as good a comparator as comparing with Birmingham or Manchester or Sheffield when they are not paying extortionate rates for agency [nurses] and things as we are here . . . When you get down to single speciality trusts [hospitals] like ours, you are narrowing down the comparator all of the time. (Interview: finance director, a London NHS hospital)

Other hospitals (although not specialist tertiary referral centres) consider that their casemix makes them worthy of special consideration. For example, a specialist or teaching hospital may take on more complex cases than those dealt with by a general district hospital within the same procedure code and/or HRG, as one finance director notes: 'The complexity of procedures is not measured in HRGs. For example, complex hip revisions undertaken at [our hospital] are referred by other orthopaedic surgeons across the country because they are too complex to be dealt with in a local district general hospital' (questionnaire comment: NHS hospital finance director).

Aside from specialist referrals, inherent characteristics of a hospital's own local population (such as social deprivation and/or demographic characteristics) can have an impact on casemix at even non-specialist hospitals. This reduces the comparability of HRGs between hospitals:

> All our old people are extremely old and amongst our young people, well there is a lot of social deprivation and drugs. Our patients may take longer to recover perhaps, on average. That results in higher costs. And, for example, our oral surgery is difficult because there is not an infrastructure here. In other areas, there are a good number of dentists that do a lot of the oral surgery. That just doesn't exist here [so] we are doing lots of things that other places aren't doing. That is going to be very difficult to adjust, and I would imagine intuitively that it's affecting our reference costs. (Interview: clinical director, a South East NHS hospital)

The most obvious aspect of determining cost efficiency is assessing the costs incurred by a hospital in running its facilities and providing health care to its patients. NHS personnel noted that substantive differences do exist in the costs incurred by different hospitals. In particular, the age, sophistication and location of a hospital's facilities influence fixed costs such as capital charges and depreciation. One clinical director expressed frustration at the distorting effect he considered such costs to have on the Index results:

> We are a new hospital. The capital cost of that hospital is added to our patient care costs. This automatically means we are 10 per cent more expensive. Why the hell is that included in the reference costs? Why do we have to pay for the fabric of the building? We could have a really crap Victorian hospital that patients loathe, and have cockroaches on the floor, and our reference costs would be 10 per cent less for exactly what we do now. Where is the logic in that? (Interview: clinical director, a South East NHS hospital)

Fixed costs are not the only contentious issue. Some direct costs, in particular labour costs, are geographically dissimilar, with London and

South East hospitals generally incurring higher costs than other hospitals (see the discussion above). In compiling the NRCE index, the NHS executive has attempted to allow for these inherent cost differences, by adjusting hospitals' index ratings using a 'market forces factor' to 'eliminate the effect of unavoidable cost differences due to different geographical locations' (DoH, 1999c, p.19). However, it is not clear that this adjustment is effective in eliminating 'unavoidable cost differences' from NRCE data. One interviewee noted: 'There is some allowance made for the fact that we're in London, but it's not enough. Our property capital values are astronomical, our staff costs are high, and I don't think the adjustment really reflects it fully' (interview: deputy finance director, a London hospital).

The fact that such inherent cost differences exist, and are so difficult to adjust for, suggests that they must have an impact on reference cost variability. The important issue here is whether such inherent cost differences are indicative of efficiency. In the long term, it might be possible for some NHS hospitals to rationalize their services or move to more cost-effective locations, but the potential for reducing many costs in the short term is clearly limited.

All of these instances of 'exceptional circumstances' imply that costs for one hospital cannot easily be interpreted alongside those for other hospitals, therefore reducing the transparency and comparability aspects of the NRCE index, and problematizing its usefulness for relative performance evaluation purposes.

DISCUSSION AND CONCLUSIONS

Financial management and costing initiatives in health care have been beset with diverse difficulties ranging from informational inadequacies to computational problems to resistance from clinicians (see, for example, Pollitt et al., 1988; Preston et al., 1992; Armstrong, 1993; Abernethy and Stoelwinder, 1995; Jacobs, 1995; Abernethy, 1996; Panozzo, 1998; Doolin, 1999). The efforts of the UK government to provide a new cost control perspective, by setting up a system of external benchmarking to highlight relative cost inefficiency, are therefore understandable (Nutley and Smith, 1998). However, the findings of this study suggest that this exercise has not, so far, been an altogether workable initiative.

A prime problem lies in identifying where benchmarking 'standards' for the Index lie. This study revealed that the NRCE initiative had failed to identify either a benchmark for excellence and the sharing of best practice or a standard of acceptable cost efficiency. Further consideration needs to be given to locating NRCE targets. In the absence of these, the 'target' equates to a moving, 'default' norm of the 100 per cent index average. The meaning of achieving 'average' cost efficiency remains undefined, yet at present this is what the hospitals are aiming for (see Llewellyn and Northcott, forthcoming). An important further question is how cost efficiency measures are to be balanced against indicators of quality, service performance and technological innovation. The resolution of this issue is fundamental to establishing an appropriate future congruence between NHS managerial controls and professional-based clinical autonomy (Jones, 1999), so that appropriate quality of care can be enabled and innovation encouraged, albeit within a context of financial constraint.

In addition to the lack of convincing benchmark standards, there is the difficulty of incommensurability between the institutions being compared, despite the attempts to deal with this through the construction of HRGs (which group together treatments that are clinically similar and consume similar amounts of resources). Moreover costing practices for these HRGs in the hospitals that feature in the Index are not sufficiently standardized to ensure confidence that varying costs really reflect different levels of efficiency rather than different cost allocation methods. The Index fails to exhibit the robustness necessary to inspire enthusiasm in its users, particularly those who define themselves as exceptional.

All of these difficulties negate the achievement of the central motifs of benchmarking: continuous technological improvement, measurement against a referent other, and rigour (Wolfram Cox et al., 1997). Measurement against a referent other is not possible without the identification of convincing 'referent others'. Rigour is denied by the lack of commensurability between the hospitals concerned and the absence of standardization in costing practices. The participants in this study desired continuous improvement, but the Index was not yet thought to be enabling improvement in any substantive way. Knowing one's position on the reference cost 'ladder of success' does little to enable a better performance without knowledge of how the 'top' hospitals achieved their results (Holloway et al., 1999). When widespread scepticism about the meaningfulness of what is being measured compounds this lack of knowledge – as here – health care improvements through reference to the Index look to be some way off. Indeed there is evidence from this study to support Allio's (1993) contention that benchmarking to a peer reference group may induce complacency rather than continuous improvement where performance is close to the average. Clearly the Index is not thought to be contributing to any enhancement in health care, as yet.

In terms of the motivations for benchmarking (improvement, consolidation, compliance and learning/teaching/mentoring), the following comments can be made. Improvement remains a motivator for government and there is no sign of any abandonment of reference costs; indeed they continue to be rolled out across all secondary and specialist health care services (DoH, 2000; p. 1). Hopes for teaching/learning/mentoring on the basis of the Index again continue to be a part of the government's agenda and there are some signs of hospital managers engaging in information sharing to this end. As yet, with the rewards for good performances uncertain, consolidation (where benchmarkees attempt to promote their relative status or superiority: Wolfram Cox et al., 1997) is not apparent. Compliance is the dominant current motivation for the hospitals' involvement in benchmarking. This conclusion supports Bowerman et al.'s (2001) argument that benchmarking in the public sector has, so far, assumed a largely defensive mode as public services acquit their newfound accountability obligations.

In an era when the government is increasingly relying on performance rankings in order to develop 'a new public sector enterprise culture in the NHS to ratchet up performance' (Health Secretary, in DoH, 2001a, p. 1),[5] the Index occupies a leading place in performance benchmarking. The findings of this study indicate that significant legitimate concerns exist regarding the use of the 'ladder of success' in a 'naming and shaming' competitive context. And the Index does provide NHS hospital managers with comparative cost information against which they can benchmark their own hospital's results and from which they can launch more detailed investigations. Can the Index be developed so as to overcome some of its associated difficulties, while retaining its political attraction as a focus for 'naming and shaming'?

The earlier sections are suggestive of several ways in which the Index could be made more meaningful (in the sense of producing more comparable data on the hospitals' cost performances). First, a distinction could be made between direct and indirect costs. Comparisons on a direct cost basis alone would avoid the problems associated with cost allocation. Also a focus on direct costs would recognize that, if there is a potential for hospitals to reduce their costs, then in the short term only direct costs are likely to be controllable.

Second, benchmarking could be focused on carefully identified comparable 'clusters' of hospitals with similar clinical/geographic/cost characteristics. This cluster approach was encouraged in earlier NHS executive publications on reference costs, but has not yet been employed within any structured framework. This approach would minimize problems relating to differential casemix and the varied clinical profiles of hospitals.

Third, further consideration should be given to identifying reference cost target outcomes for hospitals (or clusters of hospitals), rather than having a moving 'default' target of the 100 per cent index average. NHS managers recognize that being average does not necessarily indicate good performance.

Targets would best be articulated as ranges rather than absolute measures, to allow for some inherent variability in approach.

Fourth, the comprehensive scope of the Index across clinical activities could be narrowed and the ambitious government target of including all services by 2004 abandoned, in recognition that accurate costing is inherently more difficult for certain areas of activity (for example, outpatients).

However, it seems doubtful that any of these measures will be introduced, as they would result in either multiple indices, incomplete coverage or a range of target measures. If any of these were adopted, the power of a single comprehensive measure that attaches one – and only one – number to each hospital would be lost. The issue is that, in making the Index more meaningful, political leverage over the hospitals would be reduced. Hence, despite all its associated problems, the Index looks likely to continue in its present form.

This predicted prognosis is unfortunate with respect to using the Index to drive innovation in health care. The public sector, in general, has not been fertile ground for innovation (Borins, 2001) and opportunities, such as the reference cost exercise, that may provide useful comparative information for service improvement are to be welcomed. Although there is evidence to suggest that a focus on cost efficiency may inhibit innovation (Salaman and Storey, 2002), equally, pressures towards enhanced productivity may encourage managers to seek new and better ways of working. Innovation has two aspects, technical (in health care, new treatments and services) and processual (new ways of delivering services, new modes of serving patients and new types of collaborative working). In terms of process innovation, in particular, the Index could constitute a valuable conduit for information flows. It seems that the political objective of holding individual hospitals to account on the basis of a single indicator is inhibiting the potential of the reference cost exercise to enable comparable groups of hospitals to exchange meaningful data on technological development and innovation.

NOTES

* This chapter is based on 'The "Ladder of Success" in Health care: The UK National Reference Costing Exercise', *Management Accounting Research*, March, copyright 2003, with permission from Elsevier Science. The authors gratefully acknowledge the financial support of the Chartered Institute of Management Accountants in carrying out the project on which the chapter is based.

1. The NRCE will be extended to include other NHS hospitals, such as those engaged in community care and mental health services. To date, however, the majority of its data relates to acute hospitals.
2. The National Casemix Office has since been renamed the Casemix Programme.
3. Mainly health authorities and primary care trusts.

4. In February 2002, around 100 UK hospitals used a private consultancy in this way.
5. The Health Secretary, Alan Milburn, made this comment in regard to the publication of the latest NHS benchmarking initiative, a new performance ratings system that awards 'stars' to hospitals according to their performance in regard to a range of indicators (DoH, 2001b).

REFERENCES

Abernethy, M.A. (1996), 'Physicians and resource management: the role of accounting and non-accounting controls', *Financial Accountability and Management*, **12**(2), 141–56.

Abernethy, M.A. and J.U. Stoelwinder (1995), 'The role of professional control in the management of complex organisations', *Accounting, Organizations and Society*, **20**(1), 1–17.

Allio, M.K. (1993), 'The argument against adopting a process mentality', *Planning Review*, **21**, 50–51.

Armstrong, D. (1993), 'The medical division of labor', in E. Messer-Davidow, D.R. Shumway and D.J. Sylvan (eds), *Knowledges: Historical and Critical Studies in Disciplinarity*, London and Charlottesville: University of Virginia.

Bates, K. and T.J. Brignall (1993), 'Rationality, politics and healthcare costing', *Financial Accountability & Management*, **9**(1), 27–44.

Bendell, T., L. Boulter and J. Kelly (1993), *Benchmarking for Competitive Advantage*, London: Pitman Publishing.

Borins, S. (2001), 'Encouraging innovation in the public sector', *Journal of Intellectual Capital*, **2**(2), 310–19.

Bourn, M. and M. Ezzamel (1986a), 'Costing and budgeting in the National Health Service', *Financial Accountability and Management*, **2**(1), 53–71.

Bourn, M. and M. Ezzamel (1986b), 'Organisational culture in hospitals in the National Health Service', *Financial Accountability and Management*, **2**(3), 203–25.

Bowerman, M., A. Ball and G. Francis (2001), 'Benchmarking as a tool of modernisation of local government', *Financial Accountability and Management*, **17**(4), 321–9.

Broadbent, J. (1992), 'Change in organisations: a case study of the use of accounting information in the NHS', *British Accounting Review*, **24**(4), 343–67.

Broadbent, J., R.C. Laughlin and S. Read (1991), 'Recent financial and administrative changes in the NHS: a critical theory analysis', *Critical Perspectives on Accounting*, **2**(1), 1–29.

Camp, R.C. (1989), 'Benchmarking: the search for best practices that lead to superior performance. Part 1: benchmarking defined', *Quality Progress*, **22**(1), 61–8.

Department of Health (1997), 'Costed HRG Database and the National Schedule of Reference Costs', Circular FDL (97) 43, NHS executive, Leeds, December.

Department of Health (1998a), 'The new NHS – modern and dependable: a national framework for assessing performance', NHS executive, Leeds, January.

Department of Health (1998b), 'Reference cost: consultation document', NHS executive, Leeds, June.

Department of Health (1998c), 'The new NHS 1998 reference costs', NHS executive, Leeds, November.

Department of Health (1999a), 'The NHS performance assessment framework', catalogue number 16431.

Department of Health (1999b), 'The new NHS costing manual', NHS executive, Leeds, January.

Department of Health (1999c), 'The new NHS 1999 reference costs', NHS executive, Leeds, December.

Department of Health (2000), 'Reference costs 2000', NHS executive, Leeds, November.

Department of Health (2001a), press release: reference 2001/0440 (available on the DoH website http://tap.ccta.gov.uk/doh/intpress.nsf/page2001-0440).

Department of Health (2001b), 'NHS performance ratings: acute trusts 2000/01' (available on the DoH website http://www.doh.gov.uk/performanceratings/index.html).

Doolin, B. (1999), 'Casemix management in a New Zealand hospital: rationalisation and resistance', *Financial Accountability and Management*, **15**(3–4), 397–417.

Ellwood, S. (1996a), 'Full-cost pricing rules within the National Health Service internal market – accounting choices and the achievement of productive efficiency', *Management Accounting Research*, **7**, 25–51.

Ellwood, S. (1996b), 'The development of cost accounting methods in pricing healthcare', in M. Bourn and C. Sutcliffe (eds), *Management Accounting in Healthcare*, London: The Chartered Institute of Management Accountants, pp. 9–27.

Ellwood, S. (2000), 'The NHS financial manager in 2010', *Public Money and Management*, **20**(1), 25–30.

Fitzpatrick, M. and A. Huczynski (1990), 'Applying the benchmarking approach to absence control', *Leadership and Organizational Development*, **11**(9), 22–6.

Harrison, S. and C. Pollitt (1994), *Controlling Health Professionals*, Buckingham, UK: The Open University Press.

Holloway, J., M. Hinton, G. Francis and D. Mayle (1999), *Identifying Best Practice in Benchmarking*, London: The Chartered Institute of Management Accountants.

Hood, C. (1995), 'The "New Public Management" in the 1980s: variations on a theme', *Accounting, Organizations and Society*, **20**(2/3), 93–109.

Jacobs, K. (1995), 'Budgets: a medium of organisational transformation', *Management Accounting Research*, **6**(1) 59–77.

Jones, C.S. (1999), 'Developing financial accountability in British acute hospitals', *Financial Accountability and Management*, **15**(1), 1–20.

Jones, C.S. and L.P. Dewing (1997), 'The attitudes of NHS clinicians and medical managers towards changes in accounting controls', *Financial Accountability and Management*, **13**(3), 261–80.

Keen, J. (1999), 'NHS networks: rational technology or triffid?', in J. Appleby and A. Harrison (eds), *Health Care UK: The King's Fund Review of Health Policy*, London: Kings Fund, pp. 111–14.

King, M., I. Lapsley, F. Mitchell and J. Moyes (1994), 'Costing needs and practices in a changing environment: the potential for ABC in the NHS', *Financial Accountability and Management*, **10**(2), 143–60.

Klein, R. (1999), 'Has the NHS a future?', in J. Appleby and A. Harrison (eds), *Health Care UK: The King's Fund Review of Health Policy*, London: The Kings Fund, pp. 1–5.

Llewellyn, S. (1998), 'Boundary work: costing and caring in the social services', *Accounting, Organizations and Society*, **23**(1), 23–47.

Llewellyn, S. and D. Northcott (forthcoming), 'The average hospital', *Accounting, Organizations and Society.*

Mann, L., D. Samson and D. Dow (1998), 'A field experiment on the effects of benchmarking and goal setting on company sales performance', *Journal of Management*, **24**(1), 73–96.

NCMO (1997a), 'A baseline survey: healthcare resource groups (Final Report)', National Casemix Office.

NCMO (1997b), 'Uses and users of healthcare resource groups', National Casemix Office.

Northcott, D. and S. Llewellyn (forthcoming), 'Benchmarking in healthcare: exploring cost variability', Chartered Institute of Management Accountants.

Nutley, S. and P.C. Smith (1998), 'League tables for performance improvement in healthcare', *Journal of Health Services Research and Policy*, **3**(1), 50–57.

Panozzo, F. (1998), 'Clinical budgeting in Italy', paper presented at European Accounting Association Annual Congress, Antwerp.

Pollitt, C., S. Harrison, D. Hunter and G. Marnoch (1988), 'The reluctant managers: clinicians and budgets in the NHS', *Financial Accountability and Management*, **4**(3), 213–33.

Preston, A.M., D.J. Cooper and R.W. Coombs (1992), 'Fabricating budgets: a study of the production of management budgeting in the NHS', *Accounting, Organizations and Society*, **17**(6), 561–93.

Salaman, G. and J. Storey (2002), 'Managers' theories about the process of innovation', *Journal of Management Studies*, **39**(2), 147–65.

The Financial Times (1998), '*List of Shame* to Feature NHS Trust Costs', 12 June.

Wolfram Cox, J., L. Mann and D. Samson (1997), 'Benchmarking as a mixed metaphor', *Journal of Management Studies*, **34**(2), 285–314.

PART FOUR

The Intermediate View

8. The geography of intangibles: the case of the health industry

Marco R. Di Tommaso, Daniele Paci and Stuart O. Schweitzer

1 INTRODUCTION

Industrial policy for the promotion of high-tech sectors is desirable under certain circumstances. In particular, as shown by Di Tommaso and Schweitzer in this volume, this could be the case for health care-related sectors, which are usually characterized by advanced technological content. Furthermore traditional industrial policy objectives applied to health sectors can be useful in the light of a new approach to health care policy.

According to Di Tommaso and Schweitzer (2000), traditional health policies, focused only on cost containment, should be replaced by more comprehensive policies, considering health care expenditures not just as a cost, but as an investment that can have remarkable returns in terms of innovation, employment and trade. In this view, health care is considered as a broad industrial sector. The authors called this model the 'Health Industry Model', which is composed of three main actors: providers of health services (hospitals, clinics and other structures), financing organizations (both public and private) and the suppliers of health-related goods and devices.

Despite the difficulties of providing a unique definition of 'high-tech', pharmaceutical and biotechnology industry can be considered high-technology sectors both in terms of inputs (R&D and highly-skilled personnel) and in terms of outputs (complexity of products).[1] As Di Tommaso et al. (2005) noted, one of the most interesting features of high-tech sectors is their reliance on intangible assets (IAs). Intangible assets include non-physical sources of value for firms, such as patents, trademarks, knowledge, skills and relationships. However, it must be noted that in the recent and fast-growing literature on the topic, several definitions have been suggested and a number of aspects are still debated (such as how to measure intangibles and how to classify them).

These assets pose fundamental questions for managers, economists and policy makers. One of these questions is related to the geographical distribution of economic activities. The purpose of this chapter is to analyse the role played by intangible assets in health industries and whether this role is related to clustering dynamics. Intangible assets and physical proximity may appear contradictory at first sight, leading some commentators to write about 'the death of distance' with regard to the new economy and high-tech sectors.[2]

However, the analysis of health care sectors such as the pharmaceutical and biotech industries will show that these concepts are not antithetical. Rather, IAs represent a good interpretative framework to explain the locational pattern of health industries. An intermediate or 'meso' level is thus indicated as a key level of analysis for policy making in these sectors.

2 THE DEBATE ON INTANGIBLE ASSETS

The new state of the economy is usually labelled 'knowledge-based' or 'intangible' (Foray and Lundvall, 1996; Eustace, 2000), indicating that the sources of value and competitive advantage for firms are shifted from physical to intangible assets.

Simplifying, it is possible to identify three levels of analysis: the 'micro', 'macro' and the 'meso' level. At the firm's level IAs represent an increasing share of the company value and they have become the most critical factors for its competitiveness. For example, data reported by Stewart (1997) show that Pfizer spends $1.6 billion on property, plant and equipment, but nearly $3 billion on R&D. Proctor and Gamble spends $3 billion on property, plant and equipment, but $2 billion on R&D and $4 billion to promote its knowledge and build loyalty via advertising. For both Pfizer and P&G, the ratio of knowledge spending to infrastructure spending is approximately 2:1.

There is significant evidence of the importance of intangible assets also at a 'macro' level.[3] For example, according to estimates reported by Abramovitz and David (OECD, 1996) the share of tangible capital in the total stock of capital in the USA fell from 65 per cent to 46.5 per cent over the period 1929 to 1990, while the share of intangible capital[4] rose from about 35 per cent to 54 per cent as shown in Table 8.1.

A few studies have looked at intangible assets from an intermediate perspective, including work on the topic of networks and social capital[5] (Cottica and Ponti, 2004; Galassi and Mancinelli, 2004). This chapter will show that this view can be particularly meaningful in the analysis of intangibles.

Table 8.1 Capital stock and capital–output ratio in the USA, 1929–90

	1929	1948	1973	1990
Share of total capital stock %				
Tangible capital	65.1	57.8	50.2	46.5
Intangible capital	34.9	42.2	49.8	53.5
Capital–GDP ratio				
Tangible capital/GDP	7.4	6.3	5.4	5.9
Intangible capital/GDP	4.0	4.6	5.3	6.7
Total capital/GDP	11.4	10.9	10.7	12.6

Source: OECD (1996).

However, one should be careful in drawing conclusions from data and statistics reported above. Indeed one of the main problems associated with intangible assets is related to constraints in their measurement. The lack of a traditional index has consequences in terms of information asymmetry and management costs (Lev, 2001; Bianchi and Labory, 2003). The research into measuring IAs of companies has produced a plethora of proposed methods over the last few years. Four different measurement approaches can be identified:

- *direct intellectual capital methods* (DIC) consist of estimating the monetary value of intangible assets by identifying its various components. Once these components are identified, they are directly evaluated, either individually or as an aggregated coefficient;[6]
- *market capitalization methods* (MCM) are based on the calculation of the difference between a company's market capitalization and its stockholders' equity as the value of its intangible assets;[7]
- *return on assets methods* (ROA) imply that average pre-tax earnings of a company for a given period of time are divided by the average tangible assets of the company. The result is a company ROA, which is then compared to its industry average. The difference is multiplied by the company's average tangible assets to calculate the average annual earnings from intangibles. Dividing the above-average earnings by the company's average cost of capital or an interest rate, one can derive an estimate of the value of its intangible assets;[8]
- *scorecard methods* (SC) consist in identifying the various components of intangible assets and in generating indicators and indices then reported in scorecards or as graphs.[9]

Each method offers advantages and disadvantages. The first three provide a monetary estimation, while the last one is mainly focused on qualitative aspects. DIC and SC methods allow identification of every component, while MCM and ROA methods only refer to the organization level. DIS and SC methods draw a more comprehensive picture of a company's health than financial metrics, and they can be easily applied at any level of the firm.

Measurement is not the only issue in the IA literature. There is still an open debate on what exactly intangible assets are and, despite the recent claim on the theme, there is no generally accepted definition of 'intangibles'.[10] The main standard-setting accounting bodies in the world usually characterize intangible assets as identifiable (separable) non-physical and non-monetary sources of probable future economic profits accruing to the firm as a result of past events or transactions. In some cases intangibles are identified with goodwill and considered as the excess cost of an acquired company over the value of its net tangible assets (White et al., 1994).[11] However, definitions elaborated for accounting purposes may appear restrictive.

Broader definitions have been suggested in the fields of business administration and economics. Intangibles have been also defined as resources which are not visible in the balance sheet, but that contribute to add value to the enterprise (Edvinsson, 1997), or as 'non-physical sources of value (claims to future benefits), generated by innovation (discovery), unique organizational designs, or human resources practices' (Lev, 2001, p. 13). According to Eustace (2000, p. 31) intangibles are 'non-material factors that contribute to enterprise performance in the production of goods or the provision of services, or that are expected to generate future economic benefits to the entities or individuals that control their deployment'. The risk underlying these definitions is that they can be excessively broad and hence many scholars have tried to provide classifications to offer an articulated and complete picture of the complex world of intangible assets.

Lowendahl (1997), for example, categorizes intangibles into *competences* and *relational resources* (Figure 8.1). Competences represents the ability to perform a given task at both the individual and the organizational level.[12] Relational resources are, instead, reputation, client loyalty and so on. Taking the division one step further, it is possible to divide these categories into 'individual' and 'collective', depending on whether the employee or the organization is stressed. Another useful taxonomy of intangible assets is reported by Eustace (2000), who shows that the intangible constituents of the corporate asset base of modern companies can be divided into 'intangible goods' and 'intangible competences' (Figure 8.2).

The first group is made up of two main sub-classes, 'intangible commodities' and 'intellectual property'.[13] The second group, 'intangible

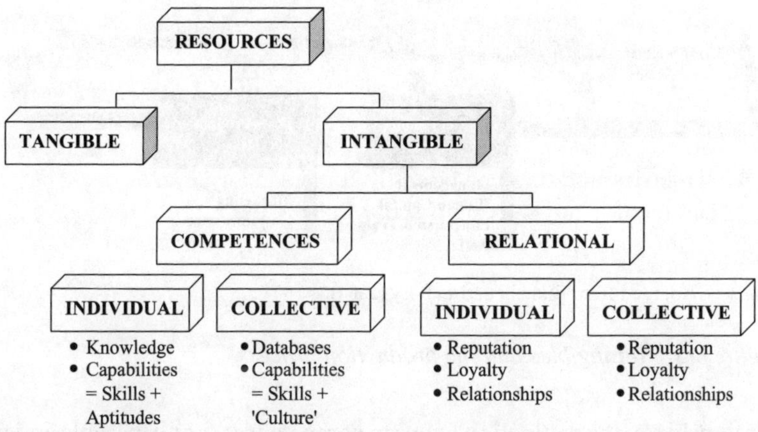

Source: Lowendahl (1997).

Figure 8.1 Intangible resources

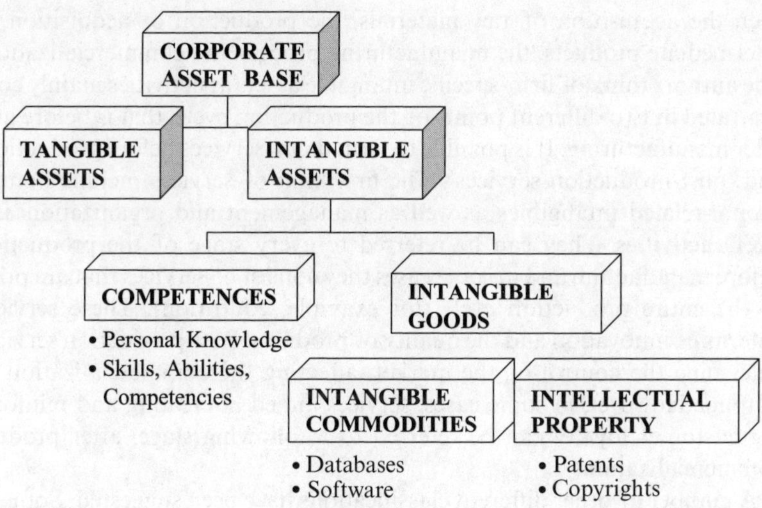

Source: Adapted from Eustace (2000).

Figure 8.2 The new corporate asset base

competences', is composed of assets that are generally bundled together and interdependent to such an extent that they are difficult (but not impossible) to isolate and value. In this case, rather than emphasize the distinction between competences (internal intangibles) and relational assets (value from

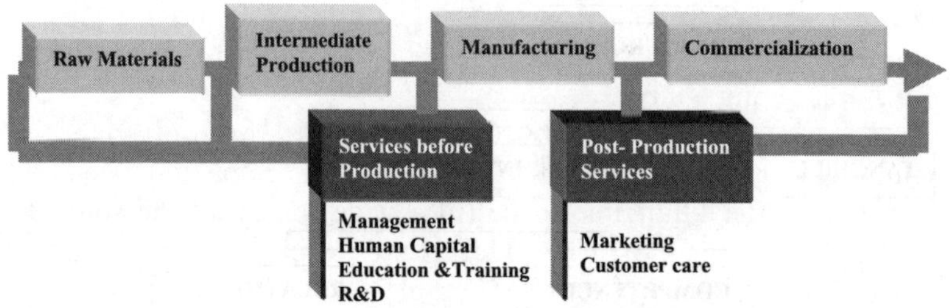

Source: Adapted from Bianchi and Labory (2004).

Figure 8.3 Intangibles and the production process

relationships), the distinction between competences (not-separable and not tradable) and intangible goods (to some extent exchangeable) is stressed.

Bianchi and Labory (2004) adopted a different perspective, considering intangible assets as services in the production process, as shown in Figure 8.3. The phases of the production cycle of a firm have traditionally been the acquisition of raw materials, the production or acquisition of intermediate products, the manufacturing process and commercialization. The authors think of firm-specific intangible assets as activities mainly concentrated in two different points of the production cycle, that is before and after manufacturing. It is possible to term them 'services before production' and 'post-production services'. The first kind of services include human capital-related intangibles as well as management and organization and R&D activities. They can be referred to every stage of the production before manufacture and in some cases they consist of services that are prior to the entire production cycle (for example, education). These services determine innovation and the quality of products. Post-production services determine the control of the market, affecting the commercialization of outputs. However, in some cases, services aimed at creating and reinforcing customer loyalty can be referred to a following stage, after product commercialization.

A number of other different classifications have been suggested. Some of them identify broad categories, while others tend to be more articulated and specific. Lev (2001) identifies three main groups of intangible assets: innovation-related intangibles, human resources and organizational intangibles. Reilly (1992) suggests that there are eight categories of intangible assets: technology-related (for example, engineering drawings), customer-related (for example, customer lists), contract-related (for example, favorable supplier contracts), data processing-related (for example, computer software), human capital-related (for example, a trained and assembled workforce), marketing-related (for example, trademarks and trade names), location-related (for example, leasehold interests) and goodwill-related (for example, going concern value).

For the purpose of this chapter it is not important to adopt one of these specific classifications. In accordance with the management and economic literature, we took intangible to be (in a wide sense) all the sources of competitive advantage for firms, which have no physical nature and are not visible in the balance sheet.

3 INTANGIBLE ASSETS IN THE HEALTH INDUSTRY

Intangible assets play a considerable role in the 'health industry'. This chapter will focus on the pharmaceutical and biotechnology industry, where value resides in brand recognition, reputation, patents and research pipelines. According to the Third Annual Knowledge Capital Scorecard developed by Baruch Lev and Mark Bothwell, and published in the April 2001 issue of *CFO*, a trade journal for financial executives, knowledge capital and intangible assets play a major role in health-related industries. Table 8.2 shows the difference in knowledge capital indexes for leading companies in different sectors.

Table 8.2 Knowledge capital estimations for selected companies, 30 Sept. 2000

Company	Knowledge capital ($ millions)	Knowledge capital/ book value	Market value/ book value	Market value/ comprehensive value
Amgen	20 786	6.00	22.40	3.20
Medimmune	4 409	6.10	24.30	3.44
Pfizer	12 861	8.60	18.20	1.90
Merck	109 217	8.60	12.60	1.32
DuPont	49 085	3.70	3.50	0.75
Microsoft	188 787	4.60	8.90	1.60
Coca-Cola	67 165	7.30	14.20	1.71
Walt Disney	53 012	2.20	3.50	1.07
Ford Motor	90 338	3.70	2.10	0.44
Exxon Mobil	114 347	1.70	4.20	1.57
Intel	208 641	5.70	13.70	2.05
Boeing	23 447	1.90	3.80	1.30

Source: CFO (2001).

Firstly Lev and Bothwell calculated the knowledge capital of each firm,[14] then each knowledge capital estimation was divided by the book value of the firm in order to measure the relative importance of knowledge capital in different companies. In Table 8.2 a comparison based on the values of the market-to-book ratio is also indicated. This shows that the stocks of several companies are traded at huge multiples of their book value (for example, Coca-Cola: 14.2; Pfizer: 18.2), but when knowledge capital is added to book value (this sum is called 'comprehensive value'), as shown in the last column, the ratio becomes more reasonable: Coca-Cola, 1.71; Pfizer, 1.90. Companies with a ratio of market value to comprehensive value significantly above one can be considered overvalued. Those with a ratio below one are probably undervalued.

Comparing the measures among firms in different sectors shows that intangible assets play a major role in biotech and pharmaceutical firms in comparison with firms in other industries. This can be seen in Figure 8.4, which compares the relative index of knowledge capital to the book value of the firms previously listed.

Looking at the industry aggregate measures of the Scorecard, the close connection between intangible assets and health industries clearly emerges. This is shown in Table 8.3.

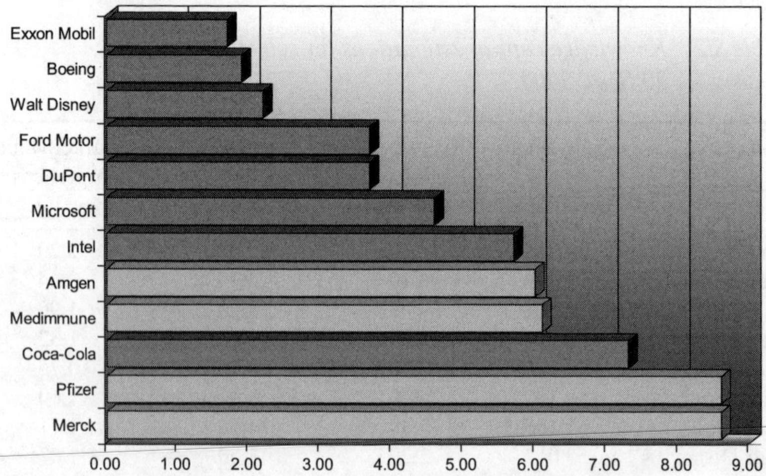

Source: Adapted from CFO (2001).

Figure 8.4 Knowledge capital/book value in selected companies, 30 Sept. 2000

Table 8.3 Knowledge capital estimations, comparisons among industries, 30 Sept. 2000

Industry	Knowledge capital ($ millions)	Knowledge capital/ book value	Market value/ book value	Market value/ comprehensive value
Aerospace & defence	23 447	3.58	1.80	0.50
Airlines	7 949	2.12	1.00	0.55
Biotech	4 393	5.18	16.30	3.07
Chemicals	9 948	3.08	2.20	0.75
Computer software	38 908	5.68	15.20	2.40
Food/beverages	18 565	7.48	9.10	1.08
Media	16 759	0.94	2.70	1.40
Motor vehicles	13 413	3.50	1.90	0.46
Oil	24 559	1.71	3.40	1.27
Pharmaceuticals	75 224	8.44	12.20	1.34
Semiconductors	42 029	2.89	3.80	1.52

Source: CFO (2001).

The results of the analysis for different industries can be seen by comparing the values of the knowledge capital/book value index.

Figure 8.5 shows that the pharmaceutical industry ranks first in the ratio of knowledge capital to book value, and that the biotech sector is close behind. This indicates the importance that intangible assets play in health industries. Another interesting result from the comparison of these indices for different industries is the importance of intangible assets in traditional sectors such as food and beverages. Even if this result depends heavily on the sample of selected firms, it is possible to argue that knowledge capital is not just a ‘new economy metric’.

Similar results can be found considering the market-to-book ratio as an index of the relevance of intangible assets (Figure 8.6). If we consider the market-to-book ratio calculated by Lev as a good proxy of the relevance of intangible assets in an industry, it is reasonable to conclude that health industries, together with the computer software industry, are the industries that are the most reliant on intangibles.[15] These results were predictable, in large part, because the pharmaceutical and biotechnology sectors are known to be based on research and innovation. In such industries, R&D expenses are notably high, as shown in Table 8.4.

Intellectual capital is crucial for health industry firms both in terms of human resources and in terms of intellectual property rights. This explains part of the importance of intangible assets within pharmaceutical and

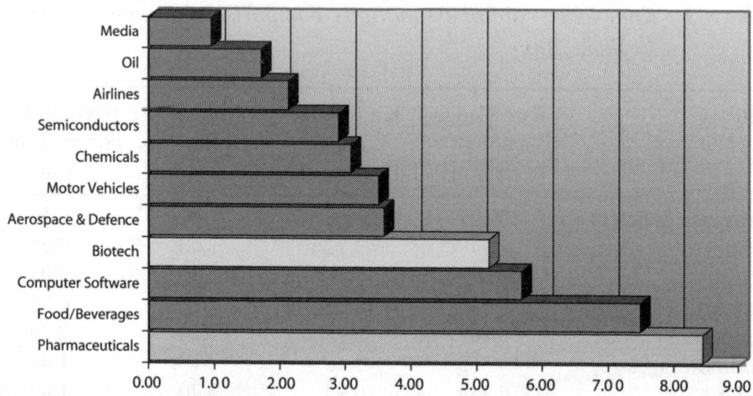

Source: Adapted from CFO (2001).

Figure 8.5 Knowledge capital/book value comparison among industries, 30 Sept. 2000

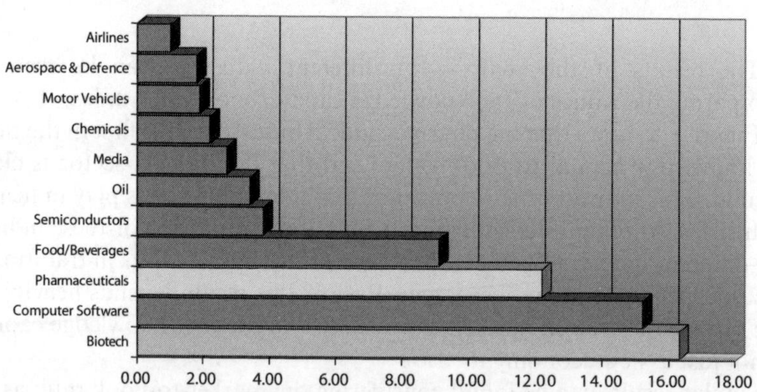

Source: Adapted from CFO (2001).

Figure 8.6 Market value/book value comparison among industries, 30 Sept. 2000

biotech firms: Pfizer spends $1.6 billion on property, plant and equipment, but nearly $3 billion on R&D (Stewart, 2001). However, other intangibles play a vital role in the health industries. Looking at the variation of the composition of costs over time (Table 8.5), it is possible to show that marketing activities have gained increasing importance in the pharmaceutical

Table 8.4 R&D indicators for pharmaceutical and biotech companies

Pharmaceutical companies	R&D expenses ($ millions)	R&D expenses per employee ($ thousands)	R&D expenses as percentage of revenue
Merck	2 119	34	6
Johnson & Johnson	2 600	27	9
Bristol-Myers Squibb	1 843	34	9
Pfizer	2 776	54	17
Eli Lilly	1 784	57	18
Biotech companies			
Amgen	823	129	24
Genentech	331	85	23
Biogen	221	164	27
Genzime	156	40	20
Chiron	254	82	37
Immunex	127	108	23

Source: Ernst & Young (2000).

Table 8.5 The changing structure of company costs in the pharmaceutical industry, 1973–89 (% of sales)

	Manufacturing	R&D	Marketing	Operating profits	Other
1973	40	10	17	23	12
1973–80	37	11	16	27	11
1989	25	15	25	29	10

Source: Jacobzone (2000).

industry. Over the 1973–89 period, the increase of marketing costs has been higher than the increase of R&D costs. The share of expenditures in marketing activities has continued to increase since the early 1990s: currently pharmaceutical companies spend, on average, 30–40 per cent of their revenue in marketing mainly to develop brands and reinforce customer loyalty.

4 INTANGIBLE ASSETS AND GEOGRAPHY

Over the 20th century, the increasing importance of non-physical inputs in comparison to raw materials, physical labour and physical capital has certainly affected the geographical distribution of economic activities. Most commentators tend to believe that recent advances in transport and communication technologies are rapidly making agglomeration economies obsolete, leading to the 'death of distance' or to a 'geographically free society' (Thurow, 1996; Cairncross, 1997; Bairstow, 2001).[16] Similar considerations can be found in less recent streams of economic theory. For example, Hall (1900) and Haig (1926) wrote that the more an industry becomes sophisticated and automated, the more its location is likely to become independent of its supply of labour and the standardized and geographically concentrated production tends to be relocated elsewhere, either closer to consumers or to cheaper input sources. But history teaches us that a reduction in transport costs has always tended to encourage geographical concentration rather than discourage it, and, as long as firms in an industry remain innovative, clustering forces remain significantly strong (Krugman, 1991).[17]

Intangible assets can be developed and managed internally by individual firms. In this case, the growing importance of intangible assets corresponds to the growth of headquarter functions and thus a geographical analysis should examine the location of multinational companies (MNCs). Recent literature on foreign direct investments and MNCs has stressed the importance of intangible assets (often called 'firm-specific assets') as one of the explanations of the advantageous possibility to extend the production activity across national boundaries (Caves, 1996).

However, the decline of the Fordist mode of production and the new overall tendency towards flexible specialization have made it possible for intangible 'services' to be developed into specialized industries, since the major firms were outsourcing these services (Kanter, 1989). These specialized firms operate mainly in sectors such as finance, legal work and marketing, but also in R&D-related activities.

A common characteristic of these firms is their tendency to cluster tightly, so that we can observe financial districts (for example, Wall Street and The City of London) and also clusters of specialized high-tech companies such as in the software industry (for example, Silicon Valley and Route 128) in the semiconductor industry (for example, Bangalore) or in biotechnology, as we will show in the following sections. In this chapter we want to underline in particular this aspect; thus, rather than analysing the growth and concentration of headquarter functions in large companies, we will show below why clusters and geographical proximity are still important in the 'intangible economy'.

5 AGGLOMERATION THEORIES REVISED

The earliest industrial agglomerations typically could be understood in terms of location of natural resources. In the 18th century, economic geography was drawn by the effort to find ways to ship raw materials from fixed source points to production locations where they could be combined with capital and labour to make final products. Because of difficulties in transporting raw materials,[18] industries used to cluster near the sources of those raw materials (Leamer and Storper, 2001; Di Tommaso and Schweitzer, 2005). Moreover there can be synergies among firms, which can gain economies by locating near one another. In particular the proximity to other similar, specialized and complementary firms can offer significant benefits in terms of access to (a) a local market for skills which reduce specialized labour search costs (b) a local specialized supply (ready to be used and less expensive) of raw materials, equipment and services, and (c) technical and market flows of specific knowledge (Marshall, 1890; Krugman, 1991).

In the 'intangible economy', the key input for companies is knowledge, that can be seen as a common denominator of the different kind of intangibles,[19] rather than traditional physical raw materials (Di Tommaso et al., 2005). But why do firms need to locate in close proximity to knowledge sources?

To answer this question it is necessary to note that there are remarkable differences between the concept of *information* and the concept of *knowledge* (Hayek, 1945; Antonelli, 1999). Alan Burton-Jones (1999) distinguishes (i) data – any signal which can be sent by an originator to a recipient – from (ii) information – data which are intelligible to the recipient – and from (iii) knowledge – the cumulative stock of information and skills derived from the use of information by the recipient. Therefore, while information can easily be codified and has a singular meaning and a unique interpretation, knowledge is vague, difficult to codify and often only accidentally recognized. The telecommunications revolution has rendered the marginal cost of transmitting information across space invariant, while the marginal cost of transmitting knowledge rises with distance (Audretsch and Feldman, 1996; Di Tommaso et al., 2005). The traditional distinction in literature is between *tacit* and *codified* knowledge.

Tacit knowledge is knowledge which cannot be dissociated from an individual, since it is stored in the brain of human beings, comprising convictions, abilities, talents and so on. On the other hand, codified knowledge is knowledge reduced and translated in a standard and compact format, that lowers the costs of storage, transmission and reproduction (David and Foray, 1995). Codified knowledge can be transferred over long distances and across national boundaries at low cost and it can be used by any number of people simultaneously (Foray and Lundvall, 1996). Tacit knowledge consists, instead, of highly specific pieces of technological know-how acquired with long processes of learning. Therefore it cannot be easily transferred, because it has not been stated in an explicit form. The transmission of knowledge (and in particular tacit knowledge) needs a process of codification and interpretation, and hence requires frequent contacts and interactions (more likely face-to-face interactions) of agents. The transmission of complex uncodifiable messages, require understanding and trust, is not likely to be affected by the Internet, which allows 'long distance conversations, but not handshakes' (Leamer and Storper, 2001). Although the development of information and communication technologies helped the process of codification of knowledge in standard form (Antonelli, 1999; Burton-Jones, 1999; Breschi, 2000), it is possible to recognize a persistent share of tacit knowledge. The boundary between tacit and explicit knowledge may shift, but codified and tacit knowledge are complementary and coexist in time. They are strongly intertwined in the human mind, so that some scholars are sceptical regarding this simple binary distinction (see Nightingale, 2003). Moreover tacit knowledge remains a key element in the appropriation and effective use of knowledge, especially when the innovation process is accelerating (Polanyi, 1967; Lundvall and Borras, 1999; Asheim, 2001). Therefore firms tend to locate in close proximity to knowledge sources: universities and research organizations, but also other firms, because tacit knowledge has a local dimension.[20]

Furthermore the traditional Marshallian forces of agglomeration among similar specialized firms still works in the intangible economy, where the possibility of having access to a skilled and specialized labour supply is a key factor of competitiveness. In addition, intangible assets generate significant localized knowledge spillovers. In a neoclassical view, knowledge is considered as a public good, since the producer cannot fully appropriate it (Arrow, 1962).

Not all intangible assets are public goods. In general they are characterized by non-rivalry. As Lev (2001, p. 34) pointed out, physical assets are generally 'rival assets', which means that alternative uses compete for the services of these assets. Differently intangible assets in general have extremely low opportunity costs. This is a common feature of assets characterized by large fixed (sunk) costs and negligible marginal costs. As knowledge-based assets, IAs are 'expensive to produce but cheap to reproduce' (Desrochers,

2001, p. 25). However, intangible assets are non-excludable only in certain circumstances. In general, in the case of intangible investments, the actors that made the investment will rarely be the only ones to derive benefits (Di Tommaso et al., 2005). But we must distinguish between IAs protected by property rights and IAs which are not protected. A firm owns its brands or patents, but it is not possible to affirm that it owns its workers and its human capital.[21] Therefore, for example, when a firm invests in training its employees, other companies (and society at large) will benefit from such investment when the trained employees change jobs (Lev, 2001). Intellectual property intangible assets such as patents, trademarks and copyrights are excludable assets. They are mobile across space and the knowledge they incorporate could be spread over long distances. In this sense such assets should lead, at least in principle, to lowering the importance of localization. However, problems of appropriability must be taken into account. For example, there are limitations on the effectiveness of patents in protecting new knowledge, which is not always perfectly embodied in legal rights. There are at least three basic limitations on this effectiveness: the ability of competitors to invent 'around a patent', the fact that some innovations are difficult to patent, and the fact that patents can disclose enough information to enable imitators to develop variants of the basic technology patented (Geroski, 1995).

Therefore, localized knowledge spillovers[22] are often indicated as important factors inducing innovative firms to cluster. The evidence is provided by numerous empirical studies on patent citation. By comparing the address of the new patent with the address of the previous patents cited it is possible to conclude that knowledge fluxes are geographically bounded and that patents usually generate localized spillover effects (Jaffe, 1989; Jaffe et al., 1993; Narin et al., 1997). Starting from these observations, Feldman (1994) concludes that innovation is an increasingly geographical concentrated process because it requires frequent formal and informal relationships among different actors at the local level and between firms and their environment.

The focus on knowledge is not incompatible with the traditional Marshallian theory of industrial agglomeration. Albino and Schiuma (1999) give evidence that knowledge is the key factor that underlies all the different interpretative frameworks suggested for industrial districts. Since knowledge can be considered as a productive force (Rullani, 1994), industrial districts can be conceptualized as 'cognitive laboratories' (Becattini and Rullani, 1993). At the basis of the Marshallian concept of industrial districts there was the so-called 'industrial atmosphere', which is made up of mutual knowledge and trust among clustered firms. Therefore it is possible to argue that the forces that drive clustering in the intangible economy are not so different from the traditional agglomeration economies.

However, the traditional focus on *static* external economies can be at least partially replaced by a growing attention to *dynamic* externalities.[23] Geographical concentration of firms is an important determinant of industrial dynamics and innovation. The creation of an 'industrial atmosphere', the presence of specialized services and advanced R&D infrastructures, the possibility of exchanging information and knowledge and of sharing similar experiences can contribute significantly to the increase of innovation opportunities and to a more rapid diffusion of technological advances (Breschi, 2000).[24] However, agglomeration economies are functional in promoting mainly incremental innovation through informal 'catching-up learning' (Asheim, 2001); that is, learning-by-doing and learning-by-using.[25] Furthermore it is reasonable to think that strong internal linkages among local firms can create difficulties in breaking path dependency and changing technological trajectory through radical innovations (Granovetter, 1973; Lazerson and Lorenzoni, 1999; Varaldo and Ferrucci, 2001). The continual accumulation of knowledge could lock firms into obsolete and increasingly non-competitive trajectories. In these circumstances, the collective learning process, which usually acts as a barrier to entry, turns into a barrier to exit (Bianchi, 1989). Learning from knowledge sourced externally is therefore an essential ingredient for the continued success of clusters and thus it would be desirable to strike a balance between internal coherence and openness to external actors.[26]

Another key issue related to intangible assets is their inherent risk and uncertainty (Lev, 2001; Di Tommaso et al., 2005). The generation of intangible assets, through investments in education, R&D and innovation, is a highly uncertain process because of its dynamic nature. The exchange of intangibles, even if protected by property rights, is characterized by a high degree of risk because of information asymmetry. Problems in recognizing and quantifying intangibles give rise to principal–agent tensions among actors involved in the exchange of these assets. Furthermore intangible assets management represents an extremely uncertain process because of the necessity of 'operating in the dark' (Lev, 2001), without an effective possibility of measuring and quantifying assets and performances.

These considerations have two effects with respect to firm localization and agglomeration. First, companies in the 'intangible economy' seek to reduce risks by locating near one another. Agglomeration economies in the traditional clustering literature include benefits from the possibility of sharing risks among the clustered firms.[27] In this sense, clustering can be seen as a mechanism to reduce the uncertainty faced by firms in connection with the development and use of intangibles in a rapidly changing environment (Lundvall, 1988; Camagni, 1991; Saxenian, 1994a). Furthermore firms seek principally specialized labour to reduce their competence gap, by locating near sources of knowledge (such as research centres and universities) or in clusters where the local concentration of firms in one industry generates a local market for specialized labour.

The second effect of uncertainty is linked to the financing of investments in intangibles. The possibility of effectively financing these assets is limited by their risky nature and by the complexity and difficulties in understanding the content of those processes that need to be financed. Therefore recent analysis identifies a key role of venture capital within regional innovative systems.[28] Venture capital investments are based on equity financing and hence they require a close interaction between the financer and the entrepreneur because the former is directly and actively involved in the business. Furthermore, because of information asymmetries, several informal aspects of the capitalist–entrepreneur relationship, such as mutual understanding and trust, gain a great importance. Venture capital investments, at least in the USA, appear tightly clustered in areas with established concentrations of high-technology businesses (Smith and Florida, 1998; Zook, 2002) and hence it is possible to suggest that venture capital and firms have a 'symbiotic role' in the formation of high-tech clusters (Florida and Kenney, 1988). Therefore the financing of intangibles requires spatial proximity between firms and their sources of financing because physical proximity can lower problems associated with failure in the credit market for the particular segment of intangible assets investments. Agglomeration of firms may be driven by the concentration of venture capitals. In other cases firms may find it advantageous to cluster because this gives capitalists more opportunity to enter into contact with them, because capitalists are likely to seek opportunities in regions where firms in the industries in which venture capitalists specialize are concentrated.

6 CLUSTER-SPECIFIC INTANGIBLE ASSETS

Storper (1997, p. 170) uses the term 'territorialization' for the process of territorial agglomeration where 'economic viability is rooted in assets (including practices and relations) that are not available in many other places and cannot easily or rapidly be created or imitated in places that lack them'. In this sense, geographical proximity can be seen as one of the most important *tangible* assets in the intangible economy. The shift from natural competitive advantage to knowledge and learning advantage has maintained regional clusters at the centre of the spatial organization of production (Florida, 1998). Learning and innovation can increasingly be seen as collective processes that are rarely confined within the boundaries of individual firms and that consist of social processes based on shared rules, trust and mutual understanding (Kline and Rosenberg, 1986; Lundvall, 1988; Feldman, 1994; Saxenian, 1994a; Antonelli, 1999; Boshma, 2001; Capello, 1999). Geographical proximity facilitates this kind of social interaction and tacit knowledge transmission. Clustering can thus generate extraordinary benefits in terms of transmission, acquisition and creation of intangibles.

As we stressed above, intangible assets have usually been considered as personal assets (such as skills and competencies) in the literature or as 'firm-specific assets' (human and organizational capital, brands, patents, R&D and so on). Recent literature has seen intangible assets also in a macro perspective, examining IAs at a national or super national level.

As Porter (1998, p. 78) argued, 'what happens *inside* companies is important, but clusters reveal that the immediate business environment *outside* companies play a vital role as well'. Localized creation and utilization of some non-ubiquitous factors – most notably tacit knowledge – can be viewed as a valuable regional asset. Therefore the analysis of collective intangible assets put forward by Lowendahl (1997) should go beyond the consideration of intangible assets within individual firms in order to consider clusters of interconnected firms as generators of relevant intangible assets such as trust, understanding and externalities (Di Tommaso et al., 2005).[29]

Figure 8.7 shows three possible levels of analysis with respect to intangible assets. IAs at the micro level can be divided into individual and firm-specific intangible assets. Individual intangible assets consist mainly of

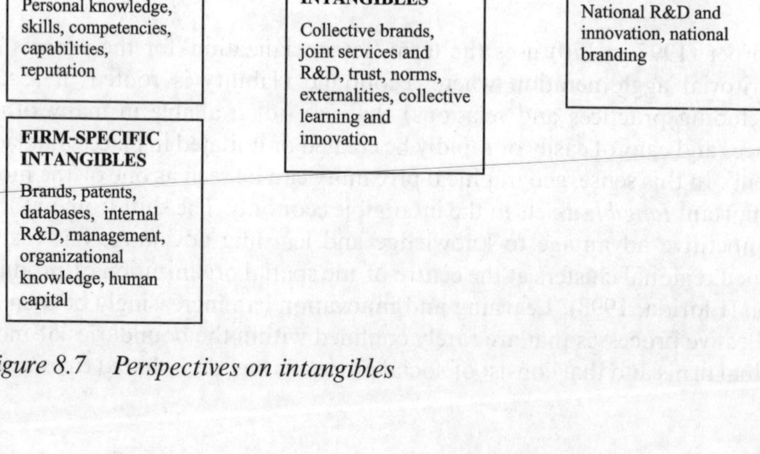

Figure 8.7 Perspectives on intangibles

competencies, personal skills and capabilities, culture, values and reputation. Firm-specific intangible assets are a combination of personal assets. They are collective assets that can be exploited by the firm as a whole. These include intellectual property rights, customer loyalty, brands, databases and human capital at the firm level. In a macro view, national (and also supernational) assets are emphasized. They can be innovation, R&D, patents and also national brands, which means the value added to a product by the fact of its being made in a specific country.

From this framework emerges an intermediate (or meso) perspective, which has been less explored in the literature so far. Cluster-specific or region-specific intangibles are collective assets that belong to a particular set of interconnected firms within the same region. These consist mainly of shared values, norms, mutual trusts and collective knowledge. In this view it is necessary to stress the importance of formal and informal institutions, which can facilitate localized collective learning through the promotion of coordination, cooperation and knowledge exchange. The organization of intangible assets through formal and informal relationships allows the increase in value of such assets. At each level of analysis, the value of intangibles is higher than the mere sum of the assets at the previous level considered separately. Therefore, as firm-specific intangible assets exceed the mere sum of the personal intangible assets in that firm, the value of cluster-specific intangible assets is higher than the sum of the assets of every single firm belonging to the cluster. However, it is important to note that this virtuous path does not always materialize and collective intangible assets may produce an obstacle to full exploitation of individual assets. This could be the case of a company that does not employ workers according to their skills or does not allow them to develop and to acquire new competences. The same can be said for clusters that cannot follow innovative trajectories because of a lack of internal coordination.

7 THE CASE OF BIOTECH CLUSTERS

In a wide sense it is possible to consider biotechnology as a collection of scientific techniques that use living cells and their molecules to make products or services.[30] However, currently, with the term 'biotechnology' scientists, researchers and public opinion indicate new technologies, developed during the 1970s and 1980s.[31] It is usual to refer to 'modern biotechnology' or 'advanced biotechnology', in order to distinguish the ancient methodologies of using living organisms (such as plant breeding and fermentation) from a set of new technologies. More specifically, we refer to biotechnology as an industry defined by specialized firms known as 'dedicated biotechnology firms' or DBFs (European Commission, 2001) that have entered the sector with the explicit aim of commercially exploiting the new technologies of life sciences.[32]

As shown previously, the biotechnology industry is one of the sectors in which intangible assets play a predominant role. This derives in large part from the science-based nature of such a peculiar sector. In fact, biotechnology is characterized by an extraordinary level of integration between industrial and scientific processes, which seems to have no comparison in other high-technology sectors (Prevezer, 1995). In the biotech industry distinctions based on the 'Schumpeterian trilogy'[33] (Stoneman, 1995), such as the distinctions between science and technology (Nelson, 1993; Stoneman, 1995) and between basic and applied research (Patel and Pavitt, 1995; Breschi, 2000), can be misleading. Biotech R&D is predominantly basic research: value is created by new ideas, rather than by new products. This has important consequences in terms of the knowledge required and produced by firms that is mainly tacit and uncodifiable. Traditionally, basic research is undertaken by universities or public institutes, because of problems of appropriability and market failure (Geroski, 1995; Breschi, 2000; McMillan et al., 2000). However, when new ideas generate extraordinary returns and opportunities, as in the case of biotechnology breakthroughs, universities can become the core of the production.[34] Therefore technology transfer mechanisms gain great importance.[35] Among them it is possible to identify transmissions via personal networks of university and industry, transfer through professional formal business relations (academic spin-offs and technology licensing) and knowledge spillovers generated by commercial applications of university physical facilities: industrial incubators, industrial parks, libraries or computer services (Varga, 1997; Pavitt, 1998).[36] Therefore access to highly-specialized human capital and to key staff is crucial. Access to graduate students and a supply of scientists and engineers represent important university–industry linkages. As a consequence, regional labour markets of scientists are responsible for much of the local academic knowledge flow (Kenney, 1986).

The biotechnology industry is predominantly composed of new small firms (Audretsch, 1999; Cooke, 2001) The relatively small scale of most biotechnology firms is arguably attributable to the 'diseconomies of scale inherent in the bureaucratic process which inhibits both innovative activity and the speed with which new inventions move through the corporate system towards the market' (Audretsch and Stephan, 1996, p. 642). In the extreme, it is possible to point out that biotechnology firms provide evidence to the theoretical prediction that, in a modern economy, increasingly characterized by environmental complexity, reliance on knowledge sources, non-price competition, vertical disintegration and higher specialization, small size may provide advantages in terms of higher competitiveness and capability to innovate (Audretsch, 1999). Interestingly the advent of small new biotechnology companies (NBCs) has not taken the place of large incumbent pharmaceutical firms (Henderson et al., 2000) and, as a consequence, in the drug market two different realities of production coexist. This happens because NBCs and established pharmaceutical companies are characterized by *complementary assets* (Malerba and Orsenigo, 2001). NBCs lack competencies in crucial aspects of the innovative process such as knowledge and experience in clinical testing and other procedures related to product approval and marketing. Thus, in a second phase, they exploited their essential competence and acted primarily as research companies and specialized suppliers of high-technology intermediate products performing contract research for and in collaboration with established pharmaceutical corporations. This provided NBCs with the financial resources necessary to fund R&D and to have access to organizational capabilities in product development and marketing. Established companies face the opposite problem. They lack the flexibility and specialization to adopt new technological paradigms quickly. Therefore, as underlined by Pammolli and Riccaboni (2000), the biopharmaceutical industry is characterized by a well articulated 'division of (innovative) labour' between large established companies and smaller biotech specialized firms.[37]

The rise of biotechnology is marked by spatial concentration: biotechnology appears as a clear example of high technology, science-based geographically concentrated industry. In the USA, biotechnology is characterized by a relatively high concentration of firms, employment and activities in a restricted number of regions, mainly San Diego, the San Francisco Bay area, Boston, Seattle, the New York metropolitan area and the Huston area (Swann et al., 1998; Zucker, Darby and Brewer, 1998; Connwell et al., forthcoming). The same phenomenon can be observed in Europe, where a small number of local clusters are capturing a dominant majority of biotechnology firms and of public research organization, such as in Paris, Cambridge, Oxfordshire, Heidelberg, Zurich, Basle and Stockholm (DTI, 1999; CENSIS, 2000; European Commission, 2001; France Biotech, 2001).

Networking and interactions among clustered firms are important factors that can enhance the innovation capacity of small biotech companies. The concept of 'collective learning' has been widely applied to biotech clusters (Cooke, 2001) and inter-firm connections are considered important not only to transfer tacit knowledge or to enjoy R&D localized spillovers, but also to enhance trust, creating social ties that are crucial for a rapid and successful adoption and diffusion of innovation. Moreover the creation of a local network of interactions enables biotech firms within the cluster to reduce the level of both technological and commercial risk, which are particularly high in this sector.[38]

Furthermore clustered firms in the biotech industry show the tendency to be 'outward looking', involving themselves in alliances with pharmaceutical and biotech companies in other regions and nations. This can be the result of marketing strategies, responding to the desire to enter new markets and possibly acquire new consumers. However, a number of agreements are based on collaboration in research and development activities for new technologies and new chemical entities. This could indicate that ties with actors external to the local area can foster innovation. As we have seen, firms with strong local ties can encounter the risk of inertia to organizational and technological change in clusters that can foster incremental innovation but limit the possibility to innovate radically. Therefore this strategy may be adopted in order to lower the risk of remaining on obsolete technological trajectories.

Different models of biotech clusters can be observed. At the risk of oversimplification, it is possible to distinguish two main typologies. The first refers principally to the USA, while the other can be considered as a model of European cases in particular.

The US Model

In the United States, the prevailing model is the agglomeration of biotech start-up companies formed by entrepreneurs and university professors, funded by venture capital (Kenney, 1986; Zucker, Darby and Brewer, 1998; Zucker, Darby and Armstrong, 1998). Therefore, in the development of biotechnology clusters, university scientists and venture capitalists have the dominant role.

The science base exerted the strongest pull on the creation of companies, which originated mainly as academic spin-offs. Therefore the geographical proximity to centres of research is crucial for these firms (Connell et al., 2003). Audretsch (1999, p. 17) states that 'biotechnology companies are defined by their scientists', identifying an active role for scientists for the development of the sector. Discoveries in this technological area are characterized by a high degree of natural excludability (Zucker, Darby and Armstrong, 1998): that is, techniques for their replication are not widely known and anyone wishing to build on new knowledge must gain access to the research team or laboratory setting having that know-how.

In these circumstances, scientists tend to enter into contractual arrangements with existing firms or start their own firm in order to extract the supernormal returns from the fruits of their intellectual contribution, and they tend to do so within commuting distance of their laboratories mainly because in numerous cases they continue to retain affiliation with their university. The first scientists who made important discoveries in the field of biotechnology entered into contact with venture capitalists, who provided financial and managerial support, enabling them to exploit their breakthroughs commercially and to appropriate the expected economic value of their tacit knowledge. Scientist-entrepreneurs[39] and venture capitalists operate in close proximity because of the significant problems of information asymmetry explained above. Therefore the local cluster involves biotech companies, universities and venture capitalists. Though networking among these actors is, in general, well developed, biotech clusters in the USA are far from being considered closed systems. In fact biotech companies tend to establish collaborations and agreements of various sorts (joint ventures, mergers and acquisitions, licensing, co-marketing agreements, R&D partnerships) with other biotech companies and other pharmaceutical companies outside their regions. In some cases they have been ready to expand into foreign markets, becoming worldwide multinationals (for example, Amgen, Chiron, Genentech).

Agglomeration of biotech firms in New Jersey represents one important exception to this model. In fact the great majority of the world's pharmaceutical companies (such as Novartis US, Aventis Pharmaceuticals, Bristol-Myers Squibb, Merck, Johnson & Johnson and Pharmacia, Hoffman-La Roche) are crowded into a narrow corridor straddling New Jersey and Pennsylvania. Established pharmaceutical companies play the key role within the local cluster: small biotech firms have strong connections with them rather than with the local academic research base.

The European Model

A process of clustering is taking place also in Europe, where biotechnology activities are concentrated in a handful of core regions. However, the development of biotechnology in Europe followed a different path. Rather than being generated by academic research, biotech companies are usually founded as corporate spin-offs from large pharmaceutical incumbents.

The weakness of university–industry relations and the scarcity of venture capital are usually indicated as two of the most important constraints to the birth of biotech spin-offs (DTI, 1999; Ernst and Young, 2001; France Biotech, 2001). In many countries the university system is only recently experiencing a radical change, introducing the possibility for scientists to exploit profitably the results of their research, for example allowing the patenting of inventions discovered in university labs. Moreover the lack of a developed venture capital market has restricted the start-up of biotechnology firms as academic spin-offs because, as previously explained, venture capitalists can provide not only finance but also managerial advice and organizational capabilities to prospective academic entrepreneurs.

Therefore, while the USA represents an example of 'market-driven' development of the biotech industry (CENSIS, 2000), in most European countries biotech has been widely promoted by specifically designed government policies such as the Bio-Regio initiative in Germany. In Europe, cases of biotech spin-offs from universities are significantly less frequent and biotech companies have been more often founded as corporate spin-offs from large chemical and pharmaceutical companies. Large pharmaceutical incumbents usually have a decisive role within the local biotech cluster because of the strong linkages between them and biotech firms. While universities and research centres cannot be considered as the engine of the development of the biotech sector, there is no doubt regarding their key role as a supportive infrastructure for the local cluster: interconnections between firms and university are frequent and involve tacit knowledge experiences and know-how not easily reproducible in a different place.

The case of the UK, where biotech firms are clustered in East Anglia (Cambridge), south-east England (Oxfordshire, Greater London, Surrey) and central Scotland, appears similar to that of the USA, because of the number of academic spin-offs and the role of venture capitalists. Most of the activities around the Oxford and Cambridge campuses as well as within the City of London are to be found within a radius of 10 kilometres. In addition to the university, Oxford includes other prestigious research organizations and hospitals (John Radcliffe Hospital, AEA Technology, the MRC Radiobiology Institute and the Wellcome Trust Human Genetics Centre). Also a number of well-known Oxford spin-offs are located along the A34 corridor from Oxford to Didcot (Oxford GlycoSciences, Oxford Asymmetry, Powderject Pharmaceuticals). Around the university campus in Cambridge are located other leading institutes (Laboratory of Molecular Biology, the Babraham Institute, the Sanger Centre and the European Bioinformatics Institute) as well as 27 per cent of UK biotech companies with a large variety of technological and business profiles (European Commission, 2001). However, it is worth noting that a large variety of actors are concentrated in the London area, including public research organizations (Imperial College, Medical Research Council, University College), research hospitals (Guy's and St Thomas's Hospital), venture capitalists and also headquarters of the main pharmaceutical and chemical enterprises.

Paris can be considered as the second-largest cluster, in terms of number of firms in Europe, after Cambridge (Mytelka and Pellegrin, 2001). Nearly 30 per cent of the French biotech firms are located in this area (France Biotech, 2000; European Commission, 2001). Another important cluster is the so-called 'Biovalley', a trinational region located in the valley of the upper Rhine, and which extends over northwest Switzerland, South Baden and Alsace. In this area are concentrated world-class research institutions (the University of Heidelberg's Centre for Molecular Biology, the European Laboratory of Molecular Biology, the German Cancer Research Centre, the Max Planck Institute for Medical Research and the Fachhochschule Mannheim) as well as some of the world's largest pharmaceutical firms (Novartis, Aventis, Johnson & Johnson, Ely Lilly, Du Pont) and chemical companies (BASF, Knoll, Boehringer Mannheim).

According to this model, local clusters include biotech companies as well as established pharmaceutical companies and research institutions. However, also in this case, as noted for the US clusters, the main clusters are characterized, not simply by dense internal or local relations, but also by the ability to establish strong and varied external ties with other firms or clusters.

CONCLUSIONS

As a consequence of the increasing importance of intangible assets, the geographical distribution of productive activities does not seem to betoken a decline of clustering and agglomeration dynamics, as some commentators have recently suggested. The globalization of markets leads to a growing localization of the sources of competitive advantage.

Industries whose value is made up of intangible assets show a clear tendency towards clustering. Numerous explanations for such a tendency can be provided. One refers to the drastic reduction of transportation costs which make it more convenient to agglomerate production in some core regions. Other explanations are linked to innovation, which is considered as an interactive process requiring the exchange of tacit knowledge, which is difficult to transfer beyond the boundaries of a local community. Furthermore the proximity between firms and financing organizations, mainly in specialized segments of the capital market, is another important consideration in the analysis of clusters in intangible assets industries.

This chapter shows that sectors of the 'Health Industry' which rely heavily on IAs represent an interesting case study: the analysis of the biotechnology industry, both in the USA and in Europe, has shown that new typologies of spatial agglomeration are emerging, challenging the ideal–typical forms of industrial district. In particular, two features of these clusters deserve mention. The first is the crucial importance of universities and other research institutions in the process of cluster formation and growth. The academic research can act as an engine of local development (as in the case of US biotech spin-offs) or as a supportive infrastructure for the local firms which gain significant advantages from different forms of technology transfer from the university (as in the case of numerous biotech clusters in Europe). The second is the tendency to establish relations external to the local cluster. These linkages are extremely important in the health industries and they can be considered as a part of knowledge capital in the sense of a set of relational resources. Such connections can be established with large established pharmaceutical companies or with other biotech companies, for several reasons. One of them, for example, is that it can be a marketing strategy aimed at acquiring new consumers in new markets, but the reason may also be found in the willingness to be connected to sources of innovation that are external to the cluster, avoiding the risk of remaining locked into obsolete technological trajectories.

Finally, while the literature on intangible assets has so far adopted a micro or a macro perspective (analysing the importance of IAs within firms or at a national or supernational level), this chapter stresses the need for a 'meso' level of analysis as a key point of view for the health industry sectors.

NOTES

1. The classifications provided in literature fall into two main groups: input-based and output-based classifications: the former identify high-tech industries as industries that have an above-average percentage of skilled personnel (engineers, researchers, computer technicians, and so on) over the total number of employees (Markusen et al., 1986) or that rely intensively on R&D (OECD, 1997; Eurostat, 2003). Output-based approaches consider the complexity or the sophistication of products (US Department of Commerce, 1982). See Malecki (1997) for a comprehensive review on the topic.
2. Cairncross (1997), Bairstow (2001).
3. See Bianchi and Labory (2004) for a comprehensive analysis on this point.
4. Intangible capital is measured by a number of proxies, including R&D spending, employment in information and communication technology, patents and public spending in education.
5. The notion of social capital is borrowed from sociology (Jacobs, 1961; Bordieu, 1986) and refers to the stock that is created when a group of organizations develops the ability to work together for mutual productive gain (Fountain, 1998). For a comprehensive review of research on this topic, see Lesser (2000).
6. This is the case of the 'technology broker' method (Brooking, 1996), according to which the value of a firm's intangibles is assessed based on diagnostic analysis of a firm's response to 20 questions covering four major components of the intangible assets.
7. Examples are the Tobin's 'q' (evocated, among others, by Stewart, 1997, and Bontis, 1999), which is the ratio of the stock market value of the firm divided by the replacement cost of its assets, and the market-to-book value (Stewart, 1997; Lev, 2001), which measures the value of intellectual capital as the difference between the firm's stock market value and the company's book value.
8. One example is provided by the economic value added (EVA™), calculated by adjusting the firm's disclosed profit with charges related to intangibles (Stewart, 1997).
9. SC methods are similar to DIC methods. The difference is that no estimate is made of the monetary value of Intangible assets. A typical example of this approach is Skandia Navigator™, which measures intellectual capital through the analysis of up to

locations. However, in many cases, especially when academic knowledge is in its evolving, non-codified stage, as in the case of biotech industry, successful knowledge transfers between university and high-technology firms require spatial proximity.

37. The recent dynamics of the market structure in the pharmaceutical industry can be summarized in two stages. In the first phase of emergence of biotechnology, a window of opportunity existed for the growth of NBCs as in Schumpeter's classic model of 'creative destruction' or 'Mark I' (Nooteboom, 2000), while thereafter the market dynamics appears oriented towards a progressive vertical integration of the production, following the model of the Schumpeterian phase of concentration (Mark II) (Mytelka and Pellegrin, 2001). It can be said that in the first phase NBCs and LECs (Large Established Countries) have entailed competitive interaction, while thereafter cooperation has prevailed. See Ernst and Young (2001) for analysis of recent mergers and acquisitions in the biotechnology market.
38. It has to be noted that risks and insecurity in biotechnology are particularly high, mainly because it is associated with a 'competence destroying' technological change (Senker, 1998) which does not allow experts to formulate any secure prediction about the future implications of new products and techniques. A comprehensive analysis of the risks connected with biotech production is provided by Buiatti (2001).
39. The emerging figure of the entrepreneurial scientist has been examined in detail by many authors. Among others, see Kenney (1986), Zucker, Darby and Brewer (1998).

REFERENCES

Albino, V. and Schiuma, G. (1999), 'New forms of knowledge creation and diffusion within industrial districts', paper presented at the Conference 'Il Futuro dei Distretti', Vicenza, 4 June.

Antonelli, C. (1999), 'The evolution of the industrial organization of the production of knowledge', *Cambridge Journal of Economics*, **23**, 243–60.

Arrow, K.J. (1962), 'Economic welfare and the allocation of resources for invention', in R.R. Nelson (ed.), *The Rate and Direction of Innovative Activity: Economic and Social Factors*, Princeton: Princeton University Press, pp. 609–25.

Asheim, B. (2001), 'Localized learning, innovation and regional clusters', in A Mariussen (ed.), *Cluster Policies – Cluster Development?*, Nordregio Report 2001:2.

Audretsch, D.B. (1999), 'Knowledge spillovers and the role of small firms', paper presented at the International Conference, 'Knowledge Spillovers and The Geography of Innovation: A Comparison of National Systems of Innovation', Saint-Etienne, 1–2 July.

Audretsch, D.B. and Feldman, M. (1996), 'R&D spillovers and the geography of innovation and production', *American Economic Review*, **86**(3), 630–40.

Audretsch, D.B. and Stephan, P.E. (1996), 'Company–scientists locational links: the case of biotechnology', *American Economic Review*, **86**(3), 641–52.

Bairstow, J. (2001), 'The death of distance', *Laser Focus World*, August, 248.

Becattini, G. and Rullani, E. (1993), 'Sistema Locale e Mercato Globale', in G. Becattini and S. Vaccà (eds), *Prospettive degli studi di Politica industriale in Italia*, Milan: Franco Angeli.

Belkaoui, M.R. (1992), *Accounting Theory*, New York: Academic Press.

Bianchi, P. (1989), 'Concorrenza dinamica, distretti industriali e interventi locali', in F. Gobbo (ed.), *Distretti e sistemi produttivi alle soglie degli anni Novanta*, Milan: Franco Angeli, pp. 65–98.

Bianchi, P. and Giordani, M.G. (1993), 'Innovation policy at the local and national levels: the case of Emilia-Romagna', *European Planning Studies*, **1**(1) 25–41.

Bianchi, P. and Labory, S. (2004), 'The economics of intangibles', in P. Bianchi and S. Labory (eds), *The Economic Importance of Intangible Assets*, Aldershot: Ashgate.

Bontis, N. (1999), 'Managing organizational knowledge by diagnosing intellectual capital: framing and advancing the state of the field', *International Journal of Technology Management*, **18**, 433–62.

Bordieu, P. (1986), 'The forms of capital', in J.C. Richardson (ed.), *Handbook of Theory and Research for the Sociology of Education*, New York: Greenwood Press, pp. 241–58.

Boschma, R. (2001), 'Proximity and innovation', paper presented at the Third Congress on Proximity 'New Growth and Territory', Paris, 13–14 December.

Breschi, S. (2000), 'La geografia delle innovazioni tecnologiche', in F. Malerba (ed.), *Economia dell'Innovazione*, Rome: Carocci, pp. 343–72.

Brooking, A. (1996): *Intellectual Capital: Core Assets for the Third Millennium Enterprise,* London: Thomson Business Press.

Buiatti, M. (2001), *Le biotecnologie. L'ingegneria genetica fra biologia, etica e mercato*, Bologna: Il Mulino.

Burton-Jones, A. (1999), *Knowledge Capitalism: Business, Work, and Learning in the New Economy*, Oxford: Oxford University Press.

Cairncross, F. (1997), *The Death of Distance*, London: Orion.

Camagni, R. (1991), 'Introduction: from the local "milieu" to innovation through cooperation and networks', in R. Camagni (ed.), *Innovation networks: Spatial perspectives*, London: Belhaven Press, pp. 1–9.

Capello, R. (1999), 'Spatial transfer of knowledge in high technology milieux: learning versus collective learning processes', *Regional Studies*, **33**, 353–65.

Caves R. (1996), *Multinational Enterprise and Economic Analysis*, Cambridge MA: Harvard University Press.

CENSIS, Forum per la Ricerca Biomedica (2000), *Nuovi pazienti e biotecnologie. Paradigmi innovativi del sistema della salute*, Milan: Francoangeli.

CFO (2001), 'CFO's third annual knowledge capital scorecard', *CFO Magazine*, April (available at the web site: http://www.cfo.com/printarticle/0,5317,2514%7CA,00.html).

Connell J., Schweitzer S.O. and Schoenberg F. (forthcoming), 'Factors underlying high-technology industrial clusters in the United States: the case of biotechnology firms', in M.R. Di Tommaso and S.O. Schweitzer (eds), *Promoting High-tech Industries. Can Health Lead the Way?*, Cheltenham, UK and Northampton, MA, USA: Edward Elgar.

Cooke, P. (2001) 'Clusters as key determinant of economic growth: the example of biotechnology' (available at the web site: http://www.nordregio.se/files/r0102cooke.pdf).

Cottica, A. and Ponti, G. (2004), 'Evolution, innovation and competition in networks', in P. Bianchi and S. Labory (eds), *The Economic Importance of Intangible Assets*, Aldershot: Ashgate.

Cowling, K. and Sugden, R. (1999), 'The wealth of localities, regions and nations; developing multinational economies', *New Political Economy*, **4**(3), 361–78.

David, P. and Foray D. (1995), 'Accessing and expanding the science and technology base', *STI Review*, **16**, OECD, Paris.

Dei Ottati, G. (1994), 'Cooperation and competition in the industrial district as an organizational model', *European Planning Studies*, **2**(2), 463–83.
Desrochers, P. (2001), 'Geographical proximity and the transmission of tacit knowledge', *The Review of Austrian Economics*, **14**(1), 25–46.
Devas, C.E. (1901), *Political Economy*, 2nd edn, London: Longmans, Green and Co.
Di Tommaso, M.R. and Schweitzer, S.O. (2000), '"The Health Industry": more than just containing costs', paper presented at the 'Health Industry Policy Forum', Ferrara, 21–22 May.
Di Tommaso, M.R., Paci, D. and Rubni, L. (2005), 'High-tech clustering: is distance dead?', mimeo.
Di Tommaso, M.R. and Schweitzer, S.O. (eds) (2005), *Health Policy and High-Tech Industrial Development: Learning from Innovation in the Health Industry*, Cheltenham, UK and Northampton, MA, USA: Edward Elgar.
Dosi, G. (1988), 'The nature of the innovative process', in G. Dosi, C. Freeman, R. Nelson, G. Silverberg, and L. Soete (eds), *Technical Change and Economic Theory*, London: Pinter, pp. 221–38.
DTI (Department of Trade and Industry, UK) (1999), 'Biotechnology Clusters', report no.1888 (available at the web site: http://biotechknowledge.com/showlibsp.php 3?uid=1).
Edvinsson, L. (1997), 'Developing intellectual capital at Skandia', *Longe Range Planning*, **30**(3), 366–73.
Edvinsson, L. and Malone, M.S. (1997), *Intellectual Capital: Realizing Your Company's True Value by Finding Its Hidden Brainpower*, New York: Harper Business.
Ernst & Young (2000), *Convergence: Ernst & Young's Biotechnology Industry Report, Millennium Edition*, New York: Ernst & Young LLP.
Ernst & Young (2001), *Integration: Ernst & Young's Eighth Annual European Life Science Report*, New York: Ernst & Young LLP.
Etzkowitz, H. and Leydesdorff, L. (2000), 'The dynamic of innovation: from National Systems and "Mode 2" to a triple helix of university–industry–government relations', *Research Policy*, **29**, 109–23.
European Commission (2001), *European Competiveness Report 2001*, Luxembourg, Office for Official Publications of the European Communities.
Eurostat, (2003), *Science and Technology in Europe, Data 1991–2001*, Luxembourg, Office for Official Pubblications of the European Communities.
Eustace, C. (2000), 'The intangible economy: impact and policy issues', report of the European High Level Expert Group on the Intangible Economy (available at the web site: http://www.euintangibles.net/publications/).
Feldman, M.P. (1994), *The Geography of Innovation*, Boston: Kluwer Academic Publishers.
Florida, R. (1998), 'Calibrating the learning region', in J. De la Mothe, and G. Paquet (eds), *Local and Regional Systems of Innovation*, Boston: Kluwer, pp. 19–28.
Florida, R. and Kenney, M. (1988), 'Venture capital, high technology and regional development', *Regional Studies*, **22**(1), 33–48.
Foray, D. and Lundvall, B.A. (eds) (1996), *Employment and Growth in the Knowledge Based Economy*, Paris: OECD.
Fountain, J.E. (1998), 'Social capital: a key enabler of innovation', in L.M. Branscomb and J.H. Keller (eds), *Investing in Innovation*, Cambridge, MA and London: MIT Press, pp. 85–111.

France Biotech (2001), 'Biotechnologies in France', available at the web site: http://www.france-biotech.org/).

Galassi, F. and Mancinelli, S. (2004), 'Why is social capital a "capital"? Public goods, co-operative efforts and the accumulation of intangible assets', in P. Bianchi and S. Labory (eds), *The Economic Importance of Intangible Assets*, Aldershot: Ashgate.

Geroski, P. (1995), 'Markets for technology: knowledge, innovation and appropriability', in P. Stoneman (ed.), *Handbook of the Economics of Innovation and Technological Change*, Oxford UK and Cambridge, MA: Blackwell, pp. 90–131.

Goldfinger, C. (2000), 'Intangible economy and financial markets', Communications & Strategies, 40, 4th quarter, 59–89.

Granovetter, M. (1973), 'The strength of weak ties', *American Journal of Sociology*, **78**(6), 1360–80.

Griliches, Z. (1995), 'R&D and productivity: econometric results and measurement issues', in P. Stoneman (ed.), *Handbook of the Economics of Innovation and Technological Change*, Oxford, UK and Cambridge MA: Blackwell, pp. 52–89.

Grossman, G.M. and Helpman, E. (1992), *Innovation and Growth in the Global Economy*, Cambridge, MA and London: MIT Press.

Haig, R.M. (1926), 'Toward an understanding of the metropolis. II. The assignment of activities to areas in urban regions', *The Quarterly Journal of Economics*, **40**(1), 402–34.

Hall, S.J. (1900), 'The localization of industries' (*Manufactures*, part 1): CXC–CCXIV, Washington: US Census.

Hayek, F.A. (1945), 'The use of knowledge in society', *American Economic Review*, **5**, 519–30.

Henderson, R., Orsenigo, L. and Pisano, G. (2000), 'The pharmaceutical industry and the revolution in molecular biology: interactions among scientific, institutional and organizational change', in D. Mowery, R. Nelson (eds), *Sources of Industrial Leadership*, Cambridge: Cambridge University Press, pp. 275–305.

Jacobs, J. (1961), *The Death and Life of Great American Cities*, New York: Random House.

Jacobzone, S. (2000), 'Pharmaceutical policies in OECD countries: reconciling social and industrial objectives', *OECD Labour Market and Social Policy – Occasional Papers*, no. 40.

Jaffe, A. (1989), 'Real effects of academic research', *American Economic Review*, **79**(5), pp. 957–70.

Jaffe, A.B., Trajtenberg M. and Henderson, R. (1993), 'Geographic localization of knowledge spillovers as evidenced by patent citations', *Quarterly Journal of Economics*, **63**(3), August, 577–98.

Kanter, R.M. (1989), *When Giants Learn to Dance*, New York: Simon and Schuster.

Kenney, M. (1986), *Biotechnology: The University–Industrial Complex*, New Haven: Yale University Press.

Kline, S. and Rosenberg, N. (1986), 'An overview of the process of innovation', in R. Landau and N. Rosenberg (eds), *The Positive Sum Strategy: Harnessing Technology for Economic Growth*, Washington, DC: National Academy Press.

Krugman, P. (1991), *Geography and Trade*, Cambridge, MA: MIT Press.

Lazerson, M.H. and Lorenzoni, G. (1999), 'The firms that feed industial districts: a return to the Italian source', *Industrial and Corporate Change*, **8**(2), June, 235–66.

Leamer, E.E. and Storper, M. (2001) 'The economic geography of the internet age', NBER Working Paper, no. W8450 (available at the web site: http://papers.nber.org/papers/W8450).

Lesser, E.L. (2000), *Knowledge and Social Capital*, Boston: Butterworth-Heinemann.

Lev, B. (2001), *Intangibles – Management, Measurement and Reporting*, New York: Brooking Institution.

Lowendahl, B. (1997), *Strategic Management of Professional Service Firm*, Copenhagen: Handelshojskolens Forlag.

Lundvall, B.Ä. (1988), 'Innovation as an interactive process: from user–producer interaction to the national system of innovation', in G. Dosi, R. Nelson, G. Silverberg, C. Freeman and L. Soete (eds), *Technical Change and Economic Theory*, London: Pinter.

Lundvall, B.Ä. and Borras, S. (1999), *The Globalising Learning Economy: Implications for Innovation Policy*, Luxembourg: Office for Official Publications of the European Communities.

Maillat, D. (1991), 'The innovative process and the role of the milieu', in E. Bergman, G. Maier and F. Tödtling (eds), *Regions Reconsidered*, London and New York: Mansell.

Malecki, E.J. (1997), *Technology and Economic Development: The Dynamics of Local, Regional and National Competitiveness*, 2nd edn, Harlow: Longman.

Malerba, F. and Orsenigo, L. (2001), 'Innovation and market structure in the dynamics of the pharmaceutical industry and biotechnology: towards a history friendly model', paper presented at the DRUID Nelson and Winter Conference, Aalborg, 12–15 June.

Markusen, A., Hall, P. and Glasmeier, A. (1986), *High Tech America*, Boston: Allen and Unwin.

Marshall, A. (1890), *Principles of Economics*, London, Macmillan; Italian version (1977), *Principi di Economia*, Turin: UTET.

McMillan, G.S., Narin, F. and Deeds, D.L. (2000), 'An analysis of the critical role of public science in innovation: the case of biotechnology', *Research Policy*, **29**(1), 1–8.

Mytelka, L.K. and Pellegrin, J. (2001), 'Can SMEs survive? Static vs dynamic externalities in the French biotechnology industry', paper presented at the 'DRUID Conference', Aalborg, 12–15 June.

Narin, F., Hamilton, K.S. and Olivastro, D. (1997), 'The increasing linkage between U.S. technology and public science', *Research Policy*, **26**, 317–30.

Nelson, R.R. (ed.) (1993), *National Innovation Systems: A Comparative Analysis*, New York: Oxford University Press.

Nightingale, P. (2003), 'If Nelson and Winter are only half right about tacit knowledge, which half? A Sarlean critique of "codification"', *Industrial and Corporate Change*, **2**(2), 149–85.

Nooteboom, B. (2000), *Learning and Innovation in Organizations and Economies*, Oxford: Oxford University Press.

Norton, R.D. (1999), 'The geography of the new economy', in R.W. Jackson (ed.), *The Web Book of Regional Science,* Regional Research Institute, West Virginia University (available at the web site: http://www.rri.wvu.edu/regscbooks.htm).

OECD (1996), *Employment and Growth in the Knowledge-Based Economy*, Paris.

OECD (1997), 'Revision of the high-technology sector and product classification', OECD STI working paper, OCDE/GD(97)216.

OECD (1999), 'Modern biotechnology and the OECD', Policy Brief, June (available at the website: http://www.oecd.org).
Pammolli, F. and Riccaboni, M. (2000), 'Technological competencies in networks of innovators: the case of biopharmaceuticals', mimeo.
Patel, P. and Pavitt, K. (1995), 'Patterns of technological activity: their measurement and interpretation', in P. Stonemann (ed.), *Handbook of the Economics of Innovation and Technological Change*, Oxford: Blackwell.
Pavitt, K. (1998), 'The social shaping of the national science base', *Research Policy*, **27**, 793–805.
Polanyi, M. (1967), *The Tacit Dimension*, London: Routledge and Kegan Paul.
Porter, M. (1998), 'Clusters and the new economics of competition', *Harvard Business Review*, November–December, 77–90.
Prevezer, M. (1995), 'The dynamics of industrial clustering in biotechnology', *Small Business Economics*, **9**, 255–71.
Reilly, R. (1992), 'Interstate intangible assets program', *CPA Journal*, **22**(8), 34–40.
Rullani, E. (1994), 'Il valore della Conoscenza', *Economia e Politica Industriale*, **82**, June, 47–74.
Saxenian, A.L. (1994), *Regional Advantage: Cultural and Competition in Silicon Valley and Route 128*, Cambridge, MA: Harvard University Press.
Schweitzer, S.O. and Di Tommaso, M.R. (2003), 'Why do biotechnology firms cluster? Some possible explanations', in R. Sugden R.H. Cheng and G.R. Meadows (eds), *Urban and Regional Prosperity in a Globalised, New Economy*, Cheltenham, UK and Northampton, MA, USA: Edward Elgar.
Senker, J. (1998), *Biotechnology and Competitive Advantage: Europe's Firms and the US Challenge*, Cheltenham, UK and Northampton, MA, USA: Edward Elgar.
Smith, D.F. and Florida, R. (1998), 'Venture capital's role in regional innovation systems: historical perspective and recent evidence', in Z.J. Acs (ed.), *Regional Innovation, Knowledge and Global Change*, London and New York: Pinter.
Stewart, T.A. (1997), *Intellectual Capital: The Wealth of Organizations*, New York: Doubleday/Currency.
Stewart, T.A. (2001), *The Wealth of Knowledge: Intelllectrcal Capital and the 21st Century Organization*, New Youk: Currency.
Stoneman, P. (ed.) (1995), *Handbook of the Economics of Innovation and Technological Change*, Oxford, UK and Cambridge, MA, USA: Blackwell.
Storper, M. (1997), *The Regional World. Territorial Development in a Global Economy*, New York and London: The Guilford Press.
Swann, G.M., Prevezer, M. and Stout, D. (1998), *The Dynamics of Industrial Clustering: International Comparisons in Computing and Biotechnology*, Oxford and New York: Oxford University Press.
Thurow, L. (1996), *The Future of Capitalism*, New York: William Morrow.
US Department of Commerce, International Trade Administration (1982), 'US high technology trade and competitiveness', September.
Varaldo, R. and Ferrucci, L. (2001), 'Cambiamenti Istituzionali nell'Impresa Distrettuale: Meccanismi Inerziali e Logiche di Evoluzione', in F. Brioschi and G. Cainelli (eds), *Diffusione e Caratteristiche dei Gruppi di Piccole e Medie Imprese nelle Aree Distrettuali*, Milan: Giuffrè Editore, 255–73.
Varga, A. (1997), 'Regional economic effects of university research: a survey' (available at the web site: http://www.rri.wvu.edu/wpapers/pdffiles/surve yattila.pdf.
Von Hippel, E. (1994), 'Sticky information and the locus of problem solving: implications for innovation', *Management Science*, **40**, 429–39.

Von Thünen, J.H. (1910), *Der isolierte Staat in Beziehung auf Landwirtschaft und Nationaloekonomie*, Jena: Fisher; English version (1966), *Isolated State*, Oxford and New York: Pergamon Press.

Weber, A. (1909), *Theory of the Location of Industries*, Chicago: University of Chicago Press.

White, G.I., Sondhi, A.C. and Fried, D. (1994), *The Analysis and Uses of Financial Statements*, New York: John Wiley and Sons.

Zook, M.A. (2002), 'Grounded capital: venture financing and the geography of the internet industry, 1994–2000', *Journal of Economic Geography*, **2**(2), 151–77.

Zucker, L.G., Darby, M.R. and Armstrong, J. (1998), 'Geographically localized knowledge: spillovers or markets?', *Economic Inquiry*, **XXXVI**, January, 65–86.

Zucker, L.G., Darby, M.R. and Brewer, M.B. (1998), 'Intellectual human capital and the birth of US biotechnology enterprises', *American Economic Review*, **88**(1), 290–306.

9. Clustering in the biotechnology industry

Stuart O. Schweitzer, Judith Connell and Fredrick P. Schoenberg

INTRODUCTION

Understanding the location decisions of firms has been one of the most important issues in industrial policy for centuries. Traditionally the field has considered the location of industrial firms in the manufacturing sector. The propensity of firms to locate near one another has been noted for many years. These earliest clusters typically could be understood in terms of location of natural resources. Steel refineries located near sources of raw materials, such as coal and iron ore, and furniture manufacturers located near sources of lumber. For other industries, access to transportation has been critical. Automobile plants located in port cities, and firms in other industries located near airports or rail or highway junctions. Still other industries sought areas where climate favoured a particular production process, or in areas that were centres of political activity.

The 'new economy' industries of the latter part of the 20th and the early 21st centuries are knowledge-based, and it would appear that the old justifications for clusters no longer apply (see Schweitzer and Di Tommaso, 2003). With output so high in terms of value per unit of size or weight, transportation costs have largely been eliminated as important considerations. Secondly, the output may not be physical at all, but rather intellectual in nature, so that 'commerce' consists more in transfer of digitized information over the internet than it does sending actual physical outputs. With these new industries, how do old predictors of clusters apply?

This chapter attempts to describe how firms in one new industry, biotechnology, decide on plant location. Armed with better information on how firms locate, governments will be better able to take measures that can encourage firms to locate in particular areas. This information will be useful to all levels of aggregation, from localities, to regions, to national governments.

BACKGROUND

The Health Industry Model (Di Tommaso and Schweitzer, 2000) describes the health system as consisting of three components: providers, payers and manufacturers. In some countries these roles are combined, especially in countries with socialized health systems combining providers and payers. The United States has a pluralistic system, with some population groups receiving care through autonomous physicians and hospitals that are reimbursed by health insurance companies. Other people are members of health maintenance organizations, which form partnerships between providers and the financing function, and still others receive care through government-owned and financed health systems. The traditional view of health systems is that their objectives were limited to the provision of an acceptable level of care at minimal cost. Systems, especially in democracies, are sensitive to demands by populations that health care be maintained at an acceptable level, but these same populations recognize that national resources are limited and so in some countries health systems are particularly frugal in terms of allocation of national income to support the health system. Other systems, particularly that of the USA, appear to reflect national tastes for higher levels of access and expenditure.

Another way of describing health systems is the extent to which they support an active research and development sector related to medical technology. Medical equipment, pharmaceuticals and biotechnology are three examples of industries that are especially prominent in some industrialized countries but are less so in others.

The Health Industry Model points out economic advantages which those industries bring to countries, in terms of scientific spillovers from one industry to another, creating wealth through expansion of high-wage and high-profit firms, and participating in the increasingly interconnected world economy in which health services, medical technology and even patients, themselves, become part of international trade (see Zucker et al., 1999; Porter, 1990; Patel, 1995). These attractions have led many industrialized countries to try to develop policies that will encourage growth of high-technology industrial sectors. In some cases the hope is that indigenous firms will start-up, and in other cases there are attempts to attract foreign firms to establish themselves in particular localities. In general, many countries are trying to create their own 'Silicon Valley' (see Farris, Hwang et al., 2001). This chapter sets out a framework to test various hypotheses concerning where high-technology firms locate. In so doing, policy instruments are described that might be useful for countries (and subnational units such as regions) to employ to attract new firms or encourage the growth of existing firms in some of these industries (see Enright, 1996).

Why do Firms Cluster?

There is an extensive literature on clustering of firms (see Krugman, 1991). Two themes present themselves. The first is that there are natural factors that draw particular firms to particular areas (Marshall, 1920). These factors may be national resources, ease of transport or location of either supply or product markets. The other theme in the literature describes synergies among firms, and describes how firms cluster together to achieve 'economies of agglomeration' (see Greis et al., 1995). These synergies may be rooted in the labour market for particular kinds of workers, the need to reduce search costs by consumers, or the desire by firms to integrate (or cooperate) either vertically or horizontally to lower costs or raise product quality. These linkages between firms seeking complementarities have been especially pronounced in Italy, and have enabled small and medium-size firms to compete globally with far larger firms. This synergy model may explain why firms follow a leader to a particular location, but it is unable to explain why the leader firm first decided to locate in a particular place. And all of these models seem inadequate to explain location of the 'new economy' industries that are particularly knowledge-intensive. Natural resources are irrelevant and, because of the peculiar nature of the products produced by high-technology firms, even transport costs are so low as to be insignificant in the production and distribution process.

One irony is that some of the industries in this 'new economy' are so well-connected by the Internet to one another, to firms that produce their inputs and to their customers that to some it seems surprising that these firms benefit at all from physical clustering (Stiglitz, 1999). One might hypothesize that these firms might function equally well if they were located at opposite ends of the Earth. And yet high-technology clusters occur throughout the world (Swann and Prevezer, 1996). Why is this? To answer this question we look first at the life cycle of firms in one particular industry, biotechnology.

The Life Cycle of Biotechnology Firms

One can make inferences about the location of biotechnology firms if we understand how these firms are often created. Biotechnology firms are frequently started as spin-offs from universities (Audretsch and Stephan, 1996). Biotech start-ups represent the combined talents of a scientist, a source of capital and management expertise. It is common, especially in the USA, for a venture capitalist literally to ask academic scientists if they have any ideas that are ready for product development. If so, negotiations are held that may result in the creation of a new enterprise. With this scenario, it is not unusual for the newly created firm to be located close to the university where the scientist continues his or her faculty association.

This scenario illustrates several important aspects of high-technology firms. The first is that the ideas come from university or research centre laboratories. The firm represents the transition from basic research to applied research where products are developed (McMillan et al., 2000). This illustrates the difference in role between non-profit research centres where basic research is conducted, often at government expense, and for-profit enterprises where applied, developmental research is conducted, often funded by investors (Narin et al., 1997; Deeds and Hill, 1996). This approach differs from some other models of firm creation, including that of Zucker and Darby (1996), who have studied the relationship between firm creation and particular 'star' scientists.

The scenario also demonstrates the importance of capital. Frequently the principals in the firm are paid little in cash, but are paid mainly in equity interest in the venture. Other costs, however, are real and must be met by actual cash. Venture capital and private placement funds are useful mechanisms for raising this initial capital, because they tend to be non-bureaucratic and geographically mobile.

Where does the Biotech Firm Locate?

According to traditional theories of firm location, one could hypothesize that biotech firms would locate near population centres, where the labour force is most abundant. This is a sort of null-hypothesis for our analysis because it says little about the particular nature of high-tech industries and firms. It suggests merely that the firms would locate where workers are, just as other firms do, at least where there are no particular natural factors (ports, highways, natural resources and so on) altering the picture.

But high-technology firms in general, and biotech firms in particular, are different (Ernst and Young, 1998). They rely on information and uniquely skilled personnel – not a typical cross-section of worker skills. This implies that a biotech firm would locate near sources of scientifically skilled personnel, perhaps near colleges and universities, and not necessarily near population centres. This hypothesis might not be specific enough, however. Our life cycle scenario suggests that the firm's initial key employee comes from a university or research institute, and is likely to retain ties to that institution. At the beginning, the need is simple: the person must retain the academic appointment in order to continue the line of basic research and to retain a salary, while other compensation from the start-up firm is merely speculative ownership shares. The university also provides other key scientific workers in the form of graduates or even graduate students. Thus a modification of the location hypothesis would be that biotech firms locate near research-oriented universities and institutions, not near educational institutions in general.

A cursory look at other industries suggests yet another hypothesis concerning the location of firms. Many firms in some industries are spin-offs, or derivatives, of existing firms (Pisano, 1991). An example is the pharmaceutical industry, which derived from the chemical industry. In fact, prior to World War II, pharmaceuticals were little more than purified chemicals produced and sold to pharmacists who compounded them and packaged them into forms that could be taken conveniently by patients, according to physician orders. If one looks at pharmaceutical firms in the USA today, one sees that many of them are located in the mid-Atlantic states, especially New Jersey and Delaware, where the chemical industry first grew in the 19th century. Taking this as a model, one might hypothesize that biotechnology firms are mere spin-offs from pharmaceutical firms and so they would tend to be located near major pharmaceutical firms. But this model of biotech firm development fails to capture the essential differences in scientific basis and paradigm between pharmaceuticals and biotechnology (Liebeskind et al., 1996). A closer observation of biotech firms shows that they grew independent of pharmaceutical firms, though a welter of mergers in recent years has brought them together, at least in terms of corporate ownership.

Two Perspectives on Firm Location

To better understand the factors determining the location of high-technology firms, there are two perspectives that can be employed. We call the first the 'county manager' view. This is the view used by regions as they attempt to attract high-technology firms. There are various policy instruments at the disposal of a regional government. Examples include property tax forgiveness and concessions to subsidize construction costs, relaxation of planning or environmental regulations, and construction of highways, rail lines, Internet links and other utility services. All of these have been used in the past to attract firms to particular cities, counties or regions.

A second perspective on the location decision is called the 'entrepreneurial view'. This is the perspective of the scientist–entrepreneur who feels ready to begin a start-up company and is deciding where to locate the firm. The entrepreneurial view focuses on things that are crucial to the decision maker's willingness to start a spin-off enterprise such as favourable university policies and availability of capital. The two views overlap, of course, as the entrepreneur certainly needs to consider the cost of establishing a firm in a particular area, and economies that result from proximity to other resources and firms.

METHODOLOGY

To test hypotheses concerning the location of biotechnology firms, our analysis is done in two stages. First, we use multiple regression analysis to explain the degree to which firm location is dependent upon several independent variables related to the location decision. Secondly, we use simulation analysis to locate biotech firms according to various location models, and we then measure the distance between *simulated* firms (whose location is simulated according to our location models) and *actual* firms. This comparison is essentially a test of the goodness-of-fit of the regression models.

The multiple regression models relate the probability that a biotechnology firm is located in a zip code (P(firm in zip_i)) to various characteristics of that zip code: pop_i (the population in zip code I), dcu_i (the distance between zip $code_i$ and the nearest college or university), dru_i (the distance between zip $code_i$ and the nearest research university) and fru_i (the level of funding of the nearest research university to zip $code_i$).

In fitting our data to the model, we encounter a problem because many of the observations of the dependent variable (the existence of one or more biotechnology firms in a zip code) are zero. We therefore use Poisson, rather than least squares, regression (Green, 1993). The choice of Poisson regression is supported by Sen and Srivastava (1990), who suggest that 'if the dependent variable is a counted variable, it is likely to have approximately a Poisson distribution' (p. 111).[1]

DATA

The study is based on an aggregated listing of US biotechnology firms and universities, both research-funded and not. The most accessible and detailed sources were found on the Internet as described below.

The smallest geographic unit of analysis was the zip code. Zip codes showing no population in the 2000 census estimate were not included. We compiled a database of 1177 biotech firms, 1573 universities and a sub-set of 396 research universities.

Biotechnology Companies

A web search was made to locate a comprehensive listing of firms. The most complete was the 'Nature Biotechnology Directory and Buyers' Guide Online', a listing of organizations, product and service providers in the biotechnology industry in 2000. It is produced in association with Nature Publishing Group Reference, publishers of *The Biotechnology*

Directory, *The Biotechnology Guide USA* and *Nature Biotechnology*, and can be found at www.guide.nature.com.

In all the sources the companies were self-defined through membership or self-selection. The directories are broad in scope, incorporating a variety of classifications such as genetics, diagnostics and therapeutics, molecular biology, immunology and microbial products and services. In addition to active research companies, the lists include suppliers and testers used by these companies. Pharmaceutical companies were not included but no exclusions were made for subsidiary biotechnology firms. We eliminated all companies that were in the agricultural, veterinary and environmental products and services group, but we had no other exclusion criteria.

Universities

The source for names and zip codes of colleges and universities in the USA was a comprehensive listing found at the University of Texas at Austin web site (utexas.edu) which contains a list of regionally-accredited four-year US colleges and universities organized by state. Zip codes and counties not available through this list were drawn from individual institution web sites.

Research Universities

Research universities were identified in two ways: designation and research activity. Designation as a research-funded university was based on participation in the National Science Foundation (NSF) Survey of Research and Development Expenditures at Universities and Colleges, which has been conducted annually since 1972. The population of institutions surveyed in most years consisted of the 500 to 700 universities and colleges that currently had doctoral programmes in science or engineering fields, or annually conducted at least \$50 000 in separately budgeted research and development (SBRD). Separately budgeted R&D is defined as current fund expenditures designed to produce specific research outcomes and either funded by a government or private agency external to an academic institution or separately budgeted by an internal unit of an institution. These institutions have traditionally received more than 95 per cent of US academic R&D funds.

The level of research activity was the amount of individual school funding reported in the survey's 'academic institutional profiles' for fiscal year 1998. Our data are drawn from the life sciences component of the survey, which is the sum of 'Agricultural sciences', 'Biological sciences', 'Medical sciences' and 'Other, not elsewhere classified'. For our purposes, we excluded agricultural sciences. R&D expenditures at 615 qualifying universities and colleges in fiscal year 1998 totalled $12.5 billion. The range for individual institutions was $10 000 to $400 million, with a mean of $26.2 million and a median of $1 million. We defined a 'research university' as one reporting at least $20 million in SBRD during calendar year 1998, which represented the top quartile of all reporting universities.

RESULTS

Distances between biotechnology firms and each zip code's population, the closest college or university, and the closest research university, are derived from the ArcGIS computer mapping software program, utilizing the zip code of each firm and college or university. Analysis of these distances is done with SAS statistical software.

The results of the regression estimation are shown below in Table 9.1. For each parameter estimate we show the standard error, the *z*-statistic, and the *p* value. Two measures of the overall goodness of fit for the Poisson regression are calculated. The first is the Akaike Information Criterion (AIC). The lower the AIC, the better the goodness of fit of the regressions. The AIC adjusts for the number of parameters being estimated. The second measure is the R^2, the proportion of variance of the dependent variable explained by the regression. This is calculated as 1 − (residual variation/null variation).

The model estimates show that all of the variables in the model are highly significant, so that a model that is constructed using only population to predict the location of biotech firms would be omitting significant variables, and hence would be an incorrect specification. The location of a nearby college or university, the location of a nearby research university and the level of funding of a nearby research university are all important predictors of the likelihood of a biotech firm being in a particular zip code. All of the coefficients have the expected sign, with higher population increasing the likelihood that a biotech firm will be located in an area, and greater distance from a zip code to the nearest college or university, or research university both reducing the likelihood. The influence of research funding is positive. The AIC and the R^2 for the models indicate that the goodness of fit improves when proximity to colleges and universities is included in the model, and it improves still further with the inclusion of proximity of research universities. Proximity to a research university is more important than proximity to a college or university, as seen by comparing model 3 with model 2. The coefficient of research funding is highly significant, and its inclusion improves the regression fit (model 5 compared to model 3).

Table 9.1 Results of the Poisson regressions predicting the likelihood of a biotech firm in a zip code

Model	Coefficient	Std error	*z*-statistic	*p*
1. intercept	-4.241	3.908×10^{-2}	-108.52	$<2 \times 10^{-16}$
pop	1.387×10^{4}	2.784×10^{2}	49.83	$<2 \times 10^{-16}$
AIC = 11,080				
$R^2 = 0.130$				
2. intercept	-3.056	5.237×10^{-2}	-58.35	$<2 \times 10^{-16}$
pop	1.063×10^{4}	3.183×10^{2}	33.38	$<2 \times 10^{-16}$
dcu	-7.285×10^{2}	3.550×10^{-3}	-20.52	$<2 \times 10^{-16}$
AIC = 10,260				
$R^2 = 0.204$				
3. intercept	-2.680	5.400×10^{-2}	-49.31	$<2 \times 10^{-16}$
pop	9.805×10^{3}	3.200×10^{2}	30.70	$<2 \times 10^{-16}$
dru	-2.300×10^{-2}	9.773×10^{-4}	-23.92	$<2 \times 10^{-16}$
AIC = 9,846				
$R^2 = 0.242$				
4. intercept	-2.386	5.610×10^{-2}	-42.53	$<2 \times 10^{-16}$
pop	9.159×10^{3}	3.212×10^{2}	28.51	$<2 \times 10^{-16}$
dcu	-4.835×10^{-2}	3.885×10^{-3}	-12.45	$<2 \times 10^{-16}$
dru	-1.611×10^{-2}	9.248×10^{-4}	-17.42	$<2 \times 10^{-16}$
AIC = 9,650				
$R^2 = 0.259$				
5. intercept	-3.196	7.167×10^{-12}	-44.59	$<2 \times 10^{-16}$
pop	9.696×10^{3}	3.227×10^{2}	30.05	$<2 \times 10^{-16}$
dru	-2.292×10^{-2}	9.514×10^{-4}	-24.10	$<2 \times 10^{-16}$
fru	3.991×10^{-6}	3.188×10^{-7}	12.52	$<2 \times 10^{-16}$
AIC = 9,708				
$R^2 = 0.254$				
6. intercept	-2876	7.354×10^{-2}	-39.11	$<2 \times 10^{-16}$
pop	9.098×10^{3}	3.240×10^{2}	28.08	$<2 \times 10^{-16}$
dcu	-4.675×10^{-2}	3.832×10^{-3}	-12.20	$<2 \times 10^{-16}$
dru	-1.592×10^{-2}	2.250×10^{-3}	-12.60	$<2 \times 10^{-16}$
fru	5.568×10^{-6}	2.659×10^{-7}	20.94	$<2 \times 10^{-16}$
AIC = 9,613				
$R^2 = 0.270$				

Model 4 includes as independent variables both the distance to the closest college or university (*dcu*) and the distance to the nearest research university (*dru*). When *dru* enters the equation in model 4, the importance of *dcu* falls, as one sees in comparing model 4 with model 2. Not only does the coefficient of *dcu* fall, but its standard error rises and its *z*-statistic falls. This suggests that the importance of a nearby college or university, while a strong predictor of the location of a biotech firm, is – to some extent – superseded by the proximity of a research-oriented university.

A similar comparison can be made with the coefficient of the research funding at the nearest research university (*fru*) in model 5 compared with model 3. As important as *fru* is in model 5 (lowering AIC and raising R^2), the coefficient of *dru* (and its standard error and *z*-statistic) changes very little. In other words, having a university in the proximity that is classified as a research university is important (model 3), but all research universities are not the same. The *size* of that university's research programme matters to a great degree.

One might suspect that multicollinearity would be a problem, as there could be high correlation between pairs of independent variables, such as population and presence of a college or a university, or between the location of a college or university and that of a *research* university. Table 9.2 shows the variance–covariance matrix for our independent variables.

The distances between simulated and actual firms to one another are shown in Table 9.3, for each of the simulation models.

These results are consistent with our hypotheses that simulation models are able to predict better where actual biotech firms are located, as one goes from the population model (model 1) to the model that incorporates

Table 9.2 Variance–covariance matrix

	Pop	dcu	dru	fru
pop	1	0.168	0.146	−0.023
dcu	0.168	1	0.690^2	0.95
dru	0.146	0.690	1	1.113
fru	0.000	0.045	0.113	1

Table 9.3 Distance between simulated and actual firms

Simulation model	Distance (km) (sd)
1	99.67 (299)
2	73.27 (287)
3	52.69 (312)
4	51.84 (320)
5	41.89 (260)
6	41.55 (271)

proximity of colleges and universities (model 2), to the model that incorporates proximity of research universities (model 3) and, finally, to the model that incorporates level of research funding (model 4).

DISCUSSION

Any observational analyses showing associations can never directly infer causality, but, with a strong theoretical reasoning behind our results, it is reasonable to impute policy implications to our findings. Our findings suggest that biotechnology firms tend to locate in populated areas, as one might expect. This is consistent with the simple model that biotech firms locate where there is a pool of workers from which to hire. But the striking part of our analysis is that the model improves markedly when we include the influence of the proximity of colleges and universities. This is consistent with the idea that biotechnology firms do not hire so much from a general population reservoir, but rather from a pool of college students or graduates. The model is an even better predictor when we include universities that are especially rich in terms of a research environment, measured either as being one of the 'elite' research institutions (in terms of government and private extramural funding) or in terms of the level of such support.

The importance of research universities can be interpreted in two ways, according to two different models of how it is that biotechnology firms locate where they do. The first is the 'spin-off' model, which says that firms are created by some sort of splitting-off of faculty (and graduate students) from a research university. But not all biotechnology firms are university spin-offs. Some represent the location decision of new firms that have no relationship to the nearby research university, or they are the relocation of an already existing biotechnology firm. In either case, when existing firms locate near research universities, it is reasonable to hypothesize that they seek access to the highly trained labour force that is already in the area. This is a kind of analog to the old clustering model in which firms locate near some kind of natural resource. In this case the 'natural resource' is a labour force that is associated with a research-oriented university.

These findings are important for any country, region or locality that would like to develop its biotechnology industry: firms in this industry tend to be located near research-oriented universities. A policy instrument for governments is clearly suggested: a precondition to the development of a strong biotechnology industry is a strong university capacity in the basic sciences that form the foundation for product development. Because of the similarities between all knowledge-intensive industries, including not only biotechnology, but also telecoms, computers and dot-coms, it is likely that a strong research university capacity is equally important for the development of these other industries. It is possible, though not tested in this project, that these industries are synergistic with biotechnology. One test of this hypothesis is whether or not high-technology clusters tend to comprise firms in a single industry (such as biotech) or, rather, a number of knowledge-intensive industries. This is a fruitful question for further investigation.

CONCLUSIONS

Our study, though suggestive of important relationships determining the location of biotechnology firms, must be interpreted cautiously. First of all, an observational study cannot demonstrate causality. Secondly, other models should be tested alongside the models we have estimated. Even within our models, there may be important confounding variables that would change our results, if they were included.

The location of high-technology industries is different from that of more traditional manufacturing firms, and these differences suggest a different set of policy instruments that can be used by localities and national governments to attract these firms and encourage their creation and growth. Though the analysis is not complete, it is likely that we will find that the strength of a country's research-oriented universities plays a major role in determining the vitality of the country's high-technology industries. The strength of the research establishment is unlikely to be sufficient, in itself, to promote the creation of high-technology firms.

There must be a legal framework to support the interests of universities, investors, and individual scientist–entrepreneurs. And secondly, reimbursement policies must exist to encourage the substantial investment in R&D that is necessary to bring high-technology products, such as biotechnology drugs, to market.

Analysis of firm location is useful in improving our understanding of the life cycle of firms, as well as suggesting policy instruments for governments desiring to promote biotechnology or other high-technology industries.

NOTES

1. An alternative approach to dealing with a large number of zero observations is to use a two-part model, breaking the location into two observations: whether or not there are firms in a zip code and, if there are firms, how many there are. We opted for a single model that includes both the dichotomous and continuous observations. The two-part model has an additional weakness because the first part treats the presence of firms as merely a

dichotomous 'yes' or 'no' question, thereby omitting useful information on the *number* of firms that might appear in a zip code.

2. The correlation between *dcu* and *dru* falls to 0.181 for zip codes ≤50 km from a college or university (32 946 out of 41 717 zip codes), and falls still further, to 0.007, for a distance less than 10 km (12 151 zip codes out of 41 717), suggesting that the high correlation shown in Table 9.2 is the result of a large number of empty zip codes with neither a college or a university nor a research university. The correlation coefficient (ρ) between population (*pop*) and *dcu* is $\rho = -0.067$, which is consistent with the observation that there are many colleges, and even research universities, that are not located in major metropolitan areas.

REFERENCES

Audretsch, D.B. and Stephan, P.E., 'Company–scientist locational links: the case of biotechnology', *American Economic Review*, **86**(3), 1996, 641–52.

Deeds, D.L. and Hill, C.W., 'Strategic alliances and the rate of new product development: an empirical study of entrepreneurial biotechnology firms', *Journal of Business Venturing*, **11**, 1996, 41–55.

Di Tommaso, M.R. and Schweitzer, S.O., 'L'industria della salute: oltre il contenimento dei costi' ('The health industry: more than containing costs'), *L'Industria: Rivista di economia e politica industriale* (*Review of Economics and Industrial Policy*), **21**(3), July 2000, 403–26.

Enright, M.J., 'Regional clusters and economic development: a research agenda', in U.H. Staber et al. (eds), *Business Networks: Prospects for Regional Development*, Berlin: Walter de Gruyte, 1996.

Ernst and Young LLP, *New Directions 98: The Twelfth Biotechnology Industry Annual Report*, Palo Alto, CA: Ernst and Young LLP, 1998.

Farris, K.F., Hwang, V. et al., 'Heart of gold: the bioscience industry in southern California', report by the Los Angeles Regional Technology Alliance, 2001.

Green, William H., *Econometric Analysis*, 2nd edn, New York: Macmillan Publishing Co., 1993.

Greis, N.P., Dibner, M.D. and Bean, A.S., 'External partnering as a response to innovation barriers and global competition in biotechnology', *Research Policy*, **24**, 1995, 609–30.

Krugman, P., *Geography and Trade*, Cambridge, MA: MIT Press, 1991.

Liebeskind, J., Oliver, A., Zucker, L. and Brewer, M., 'Social networks, learning and flexibility: sourcing scientific knowledge in new biotechnology firms', *Organization Science*, **3**, 1996, 783–831.

Marshall, A., *Industry and Trade*, London: Macmillan, 1920.

McMillan, G.S., Narin, F. and Deeds, D., 'An analysis of the critical role of public science in innovation: the case of biotechnology', *Research Policy*, **29**, 2000, 1–8.

Narin, F., Hamilton, K. and Olivastro, D., 'The increasing linkage between US technology and public science', *Research Policy*, **26**, 1997, 317–30.

Patel, P., 'Localised production of technology for global markets', *Cambridge Journal of Economics*, **19**, 1995, 141–53.

Pisano, G., 'The governance of innovation: vertical integration and collaborative arrangements in the biotechnology industry', *Research Policy*, **20**, 1991, 237–49.

Porter, M.E., *The Competitive Advantage of Nations*, New York: Free Press, 1990.

Schweitzer, S.O. and Di Tommaso M.R., 'Why do biotechnology firms cluster? Some possible explanations', in R. Sugden, R.H. Cheng and R. Meadows (eds),

Urban and Regional Prosperity in a Globalised, New Economy, Cheltenham, UK and Northampton, MA, USA: Edward Elgar Publishers, 2003.

Sen, A. and Srivastava, M., *Regression Analysis: Theory, Methods and Applications*, New York: Springer-Verlag, 1990.

Stiglitz, J., 'Public policy for a knowledge economy', speech for Department for Trade and Industry and Centre for Economic Policy Research, London (http://www.worldbank.org/html/extdr/extme/jssp012799a.htm, 27 January 1999).

Swann, P. and Prevezer, M.A., 'A comparison of the dynamics of industrial clustering in computing and biotechnology', *Research Policy*, **25**, 1996, 1139–57.

Zucker, L.G. and Darby, M.R., 'Star scientists and institutional transformation: patterns of invention and innovation in the formation of the biotechnology industry', *Proceedings of the National Academy of Sciences*, **93**(23), 12 November, 12709–16.

Zucker, L.G., Darby, M.R. and Armstrong, J., 'Intellectual capital and the firm: the technology of geographically localized knowledge spillovers', National Bureau of Economic Research, working paper no. 4946, Cambridge, MA, 1999.

10. Spillovers of university–high-tech industry alliances

Werner Z. Hirsch

INTRODUCTION

In line with the notion of spillovers incorporated in the Health Industry Model of Di Tommaso and Schweitzer (2001), this chapter explores this phenomenon in regard to a rather novel form of collaboration between research universities and high-tech industries. It will point to the benefits that are expected to flow from research alliances, a major reason why these have become increasingly common. Modelling of the spillover process will come next, along with an analysis of the regional clustering process, to be followed by a case study of the spillover process. Finally, some important further research needed in this area is explored.

To be a player in the knowledge-based high-tech economy, often looked upon as a key element in the 'new economy', requires successful and timely innovation and product inventions for which there will be a responsive demand. In contrast to the 'old economy', the high-tech industries of the new economy, especially pharmaceuticals, semiconductors and computer software, incur extremely high development and start-up costs, inordinately low production costs and yet exceptionally rapid obsolescence. While, thus, high-tech companies have a strong incentive to collaborate with research universities, so have the universities to collaborate with industry. These mutual interests have given impetus to the formation of new forms of university–high-tech industry collaboration, with research alliances being perhaps the most prominent. However, such collaboration not only benefits a specific university and firm, but provides spillovers to other universities and firms as well as to the economy in general.

BENEFITS OF COLLABORATIVE EFFORTS

Universities and industry at first collaborated mainly on an individual basis, faculty serving as consultants, board members and part-time employees.

More recently, collaborative efforts have taken the form of university–industry research alliances. They often involve teams of faculty researchers, at times whole departments, as in the biotechnology collaboration between UC Berkeley and Novartis (sometimes government serves as a third party, as in the California Institutes for Science and Innovation). Alliances between research universities and the high-tech industry have become a hallmark of the new economy, growing rapidly in number and funding. Today, in America, alliances are funded by corporations to the tune of $2 billion a year, twice what it was a decade ago (National Science Foundation, 1998).

A major reason for forming research alliances is the self-interest of both high-tech firms and research universities. Not only do the two benefit from collaboration, but so do regional economies, as well as nations and society at large.

For universities, positive driving forces include the quest for new revenue sources and intellectual gains from collaborating in research with scientists in industry who are used to working on real-world problems, who often have vast experience and have developed a rather unique culture and way of thinking. Industry (and government laboratories) brings to the table expensive world-class equipment and instrumentation as well as financial resources. Alliances also facilitate the placing of the university's graduates.

Industry benefits, since universities bring to the table world-class scientists and a well-educated staff, as well as patents and an environment which stimulates inquiry and creativity (for example, the top 173 American universities' 1996 royalty and licence fee earnings were $592 million) (Marcus, 1998). Moreover industry benefits because outsourcing of research to universities enables it to engage the very best scientists, who often are unwilling to work in industry, and also to gain greater flexibility in manning their research efforts.

Research universities not only have a great capacity for discovering and inventing new products and processes of great value, but also own a host of patents. The latter fact has been reinforced by the passage of the Bayh–Dole Amendment, which encourages universities to license to private industry discoveries made with federal funds. For example, of the 2800 US human gene and genomic patents granted in 1999, only about 900 were granted to private firms, while 1900 went to academic institutions (*Harvard Magazine*, 2001).

Let me offer an example of the mutual benefits of collaboration: the Monsanto Co. (now Pharmacia Corporation) has funded research at Washington University in Saint Louis since 1981, the funds amounting to $100 million during that period. In turn, Washington University received

180–90 patents, but the link also led to some personnel trades. The most prominent is the faculty member who transferred in 1989 to Monsanto Co. and is now Pharmacia's chief scientific officer (Business–Higher Education Forum, 2001).

Society at large can benefit, since alliances tend to stimulate the creation of new knowledge, discoveries, innovation and inventions. These spillovers are most prolific in the presence of high-tech industry clusters, to whose growth they in turn can greatly contribute. A major result of these spillovers and clusters is a host of regional as well as national economic benefits. This expectation must have been in the mind of California Governor Gray Davis in the establishment and funding of the California Institutes of Science and Innovation, which require that, for the $300 million provided by the state, the universities must raise an additional $600 million. He said, 'It's my hope to replicate Silicon Valley . . . The most important thing a state government can do to improve local economies is to support research universities' (Markoff, 2000). Corporate funding has followed rapidly. For example, one of the institutes alone received at the outset $140 million from companies such as IBM, Sun Microsystems, Qualcom and Sony.

Regions and the nation reap economic gains when alliances produce discoveries and innovations. Also synergies from complementary integration, and productivity gains from vertical disintegration through outsourcing, as well as scale economies from horizontal integration, can be expected. Altogether research alliances can have a seedbed effect, stimulating the emergence of high-tech clusters which further raise productivity and foster innovation and economic growth.

MODELLING ECONOMIC SPILLOVERS

As was mentioned in the preceding pages, research alliances can not only benefit their partners, but can also affect the economic health of the region in which they are located, with spillovers to the rest of the state and nation. For an analysis of the effects on expenditure and employment we can apply regional impact analysis developed by economists and extend it to three stages, as presented in Figure 10.1. Thus, in stage I, we have the direct impact on the regional economy emerging from the university spending the funds of the corporate research contract on labour, material and services. Stage II reflects the indirect and income-induced effects, and stage III the seedbed effect of the research grant. All of these effects have significant geographical dimensions, so that the alliance's total impact on local and regional economies is significantly greater than the sum of direct expenditures funded by the research contract.

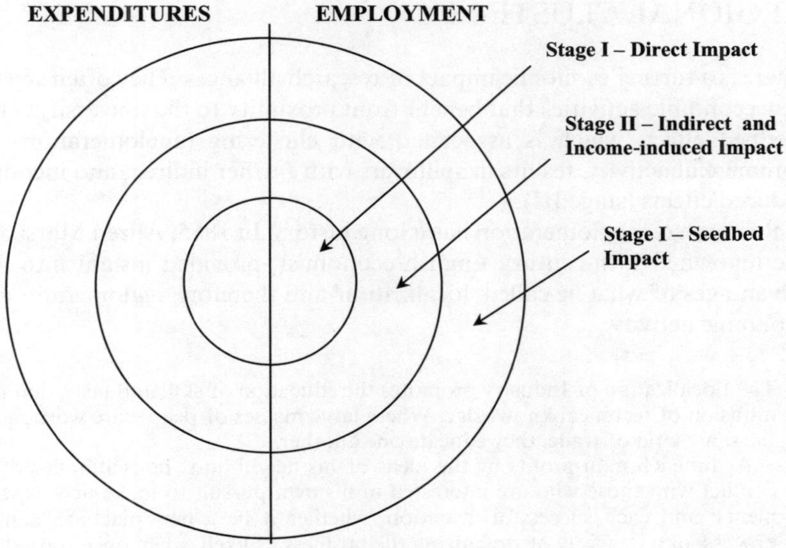

Figure 10.1 Three impact stages of university–high-tech industry research alliances

Two major interrelated forces are responsible for the regional economic impact of alliances: the inter-industry multiplier effect of dollars expended by alliances on labour, services and material which cycle through the economy several times, and the emergence of high-tech clusters which stimulate discoveries, innovation and economic growth.

Economists refer to the recycling of money spent on labour, material and services in an economy as the indirect and income-induced 'multiplier effect', so crucial in stage II. The impact of each dollar spent is 'multiplied' as it is spent again in the economy. For example, the salaries paid by the university to faculty and staff are spent by them to buy food, transport, clothing, schooling and so on. To produce these and other goods and services, producers must buy a host of inputs, including labour. The extent of the effect can be estimated by using inter-industry multipliers, which can be calculated by modelling regional economies and making econometric estimates of their magnitude. For example, a regional expenditure multiplier of 2.0 indicates that the combined direct, indirect and income-induced effect is twice as large as the initial expenditure.

REGIONAL CLUSTERS

There is a further economic impact of research alliances. They often spawn new economic activities that benefit from proximity to the university. This seedbed effect, which is associated with clustering (agglomeration) of commercial activity, results in spillovers with further indirect and income-induced effects (stage III).

Research on agglomeration has a long history. In 1885, Alfred Marshall, the renowned 19th-century English economist, provided insight into the advantages of what he called 'localization' and therefore agglomeration of economic activity:

> The Localization of Industry promotes the education of skill and taste, and the diffusion of technical knowledge. Where large masses of people are working at the same kind of trade, they educate one another.
>
> Again each man profits by the ideas of his neighbours: he is stimulated by contact with those who are interested in his own pursuit to make new experiments; and each successful invention, whether it be a new machine, a new process, or a new way of organizing the business, is likely when once started to spread and to be improved upon.
>
> In a district in which an industry is localized a skilled workman is sure of finding work to suit him; a master can easily fill a vacancy among his foremen; and generally the economy of skill can be carried further than in an isolated factory however large. Thus both large and small factories are benefited by the localization of industry and by the assistance of subsidiary trades. (Marshall and Marshall, 1885)

Thus, just as Marshall's localization effects are long-term, cumulative and depend on cooperation in knowledge creation and innovation, so does high-tech clustering. Marshall's ideas are suggestive for analysis of contemporary clustering.

Financial success of knowledge-based high-tech corporations requires timely discoveries, inventions and innovation. Firms, as indicated earlier, face, on the one hand, exceptionally high development costs and therefore high start-up costs, while production costs are extremely low, and, on the other hand, exceptionally rapid obsolescence. Pharmaceuticals, semiconductors and computer software exemplify these circumstances. For example, bringing a new drug to market can cost as much as half a billion dollars. The high cost is related to the fact that, 'for every 5,000 compounds evaluated for treatment, only five will make it to clinical trials of which just one will make it to market' (Shaywitz and Ausiello, 2001). Usually this takes many years. The same holds for semiconductors and software. At the same time, the useful life of semiconductors is about a year and a half.

Under these circumstances, the greatest rewards in knowledge-based enterprises go to companies that innovate at a rapid pace and thereafter obtain the largest possible market share for their new product. Consequently firms in high-tech industries are consumed by the defining drive to innovate and, if possible, achieve monopoly power, however temporary it turns out to be.

Financial success is greatly facilitated by collaborating with research universities and locating in their vicinity. Universities are thus increasingly surrounded by clusters of symbiotic enterprises. These clusters benefit from synergies and positive externalities on the demand side and from cost savings on the supply side. In turn, they attract human capital of the highest quality while providing an environment conducive to lively exchanges of knowledge and ideas.

A high-tech cluster is thus a geographic concentration of horizontally and vertically interconnected companies and associated institutions, which have located themselves around research universities and other research institutions. The latter two tend to be at the core, with a host of high-tech knowledge-based institutions surrounding them linked by commonalities and complementaries, while benefiting from positive externalities. Physical proximity among those who work on the cutting edge of knowledge continues to be extremely valuable, even in an age where the cyberspace revolution has shrunk distances in space and time. 'Even in the days of instantaneous communication, there is no substitute for researchers pressing flesh . . . and the ability to sit in the bar and chew the fat with colleagues and rivals' (*The Economist*, 1999).

Demand-related horizontal interactions tend to be crucial for initiating the clustering process. Benefits from these interactions include the ease and timeliness with which information, often unplanned, is exchanged between cap and gown and among high-tech firms. In addition to horizontal, demand-related forces, there also exist significant vertical, supply-related ones. As firms form clusters, they need inputs, not only scientists and staff, but also products and services, so that they can carry out their missions. This supply-related growth follows the demand-related one, but in due time they tend to interact. Being located in a high-tech cluster, and thus having access to a large labour pool and to specialized inputs, can raise a firm's productivity and competitiveness. Much of a firm's outsourcing can be local and thus involve lower transaction costs than non-local outsourcing, but only up to a point. When clusters get too large and too cluttered with enterprises, negative externalities tend to set in, raising transaction costs. Traffic congestion, air pollution and bid-up housing prices and wages are examples.

Horizontal and vertical interactions sooner or later affect each other. For example, as suppliers of inputs exchange information and ideas with high-tech research firms and universities, they in turn provide scientists and their students, and consequently improve the productivity of suppliers of goods and services. Because of these manifold interactions, technological developments, dynamics of the market and government regulation, high-tech clusters are in a continual state of flux.

ESTIMATES OF REGIONAL SPILLOVERS: A CASE STUDY

With the help of the analytic framework developed in the preceding section, rough estimates are made of the expenditure and employment effects of recent corporate research funding at the University of California, San Diego (UCSD), the University of California as a whole (UC), and all California research universities (Hirsch, 2000). The following assumptions are made.

1. There is slack in both the university and the local economy. Thus both can readily meet the demand for personnel and foods and services needed under the contracts.
2. In the absence of research contracts, no substitute would have taken their place.
3. On the basis of empirical studies of UC Irvine and UC Los Angeles, both of which spent about 80 per cent of their operating budgets in their respective surrounding counties and 95 per cent within California, these percentages were used.
4. Alliance-related annual expenditures are assumed to be the sum of corporate research grants and half the royalties and fees earned in a given year.
5. On the basis of a review of empirical expenditure multiplier studies of universities, a range with a low of 1.6 and a high of 3.0 is assumed for stage II.
6. On the basis of UC San Diego data, 17 750 jobs were associated with a $1.1 billion annual budget; that is, 16 jobs per $1 million expenditure. For UC, employment per $1 million expenditures was 11 and for all California research universities it was assumed to be 13 (the figures are used to translate stage II and stage III expenditures into employment impacts).
7. On the basis of a UC San Diego study, which found that the university's presence stimulated the founding in San Diego County of spin-off firms with annual revenue of $2 billion, while the university budget was $1.1 billion, the seedbed factor was estimated to be 1.8.

San Diego is an interesting case of high-tech (heavily biotech) clustering around a research university (and two research institutes, each no more than two miles away from UCSD – Scripps Institute and Salt Institute (Glover and Bibbens, 2000). No doubt the existence of UCSD and the two research institutes was a major factor in the National Science Foundation's 1985 decision to locate one of its five national computer centres on the UCSD campus, and in funding UCSD in 1997 as one of its two national consortia.

The emergence of San Diego as a major biotechnology cluster is no doubt the result of networking, fostered by the visionary role played by Richard Atkinson, then chancellor of UCSD. In 1984, he founded CONNECT, which developed powerful instruments to support an emerging high-tech industry in San Diego. From 17 company sponsors in 1985, CONNECT had 720 members by 2000. Of San Diego's 250 biotechnology companies, most of them start-ups, 124 were active in the CONNECT network.

With the help of these assumptions some tentative regional economic impact estimates are produced. As can be seen in Table 10.1, UC San Diego's 1998 corporate grants and half of its royalties and licence fees of \$40.4 million were found to have had a total stage I and stage II impact on the economy ranging from a minimum of \$64.6 million to a maximum of \$121.2 million.[1] Eighty per cent, or somewhere between \$51.7 million to \$97.0 million, were spent in San Diego County, while the comparable figures for California were 95 per cent and \$61.4 million to \$115.1 million. Adding the stage III impact, overall expenditures in San Diego county resulting from its alliances amounted to between \$93.1 and \$174.6 million.

Table 10.1 Economic impact of UC San Diego's corporate research support (and royalties plus fees) in 1998

	Expenditure (in US\$ m.)			Employment**			
	Total	County	State	Total	County	State	
Grants	37.4						Stage I
Royalties and fees*	3.0						
Total	40.4	32.3	38.4	646.4	517.1	614.1	
1.6 multiplier	64.6	51.7	61.4	1 033.6	826.9	981.9	Stage II
3.0 multiplier	121.2	97.0	115.1	1 939.2	1 551.4	1 842.2	
1.8 seedbed factor		93.1			1 489.6		Stage III
		174.6			2 793.6		

Notes: * Half of the year's royalties and licence fees; ** 16 jobs per \$1 million expenditure.

Turning to the employment impact of UC San Diego's alliances, we find in Table 10.1 that their corporate contracts plus half their royalties and licence fees required in stage I a total of 646.4 employees, of which 517.1 and 614.1 lived in the county and state, respectively, while the rest lived outside them. In stage II, total employment increased by between 1033.6 and 1939.2, with 80 per cent living in San Diego County and 95 per cent in California. Under the above assumptions, by including a seedbed factor, the county's employment in stage III increased to between 1489.6 and 2793.6 jobs.

The expenditure and employment impacts of UC's and all California's research universities' alliances can be found in Table 10.2. UC's corporate contracts plus half its royalties and licence fees amounted in 1998 to $213.6 million. Adding their stage II and stage III effects results in estimates of between $584.5 and $1095.8 million. The corresponding employment impact is between 6429.5 and 12 053.8 jobs, many of them held by well-educated and highly skilled personnel paid relatively high salaries.

In Table 10.3 and Figure 10.2, the impact of all the California's research universities is given (including the three private ones with corporate contracts and half their royalties and licence fees of $274.3 million). The overall expenditure impact is estimated to fall by between $750.6 and $1407.2 million and the employment impact by between 9747.8 and 18 294.1 jobs. The significance of the economic impact of California's research alliances on its economic health can be appreciated by realizing that UC San Diego's annual expenditures and employment fall within the range of the above estimates. In short, the economic impact of California's research alliances is about equal to the budget and employment of UC San Diego, one of California's major research universities.

Table 10.2 Economic impact of University of California's corporate research support (and royalties plus fees) in 1998

	Expenditures (in US$ m.)		Employment**		
	Total	State	Total	State	
Grants	173.7				Stage I
Royalties and fees*	39.9				
Total	213.6	202.9	2 349.6	2 232.1	
1.6 multiplier	341.8	324.7	3 759.8	3 571.5	Stage II
3.0 multiplier	640.8	608.8	7 048.8	6 696.4	
1.8 seedbed factor		584.5		6 429.5	Stage III
		1 095.8		12 053.8	

Notes: * Half of the year's royalties and licence fees; ** 11 jobs per $1 million expenditure.

Table 10.3 Economic impact of California's research universities' corporate research support (and royalties plus fees) in 1998

	Expenditures (in US$ m.)		Employment**		
	Total	State	Total	State	
Grants	223.0				Stage I
Royalties and fees*	51.3				
Total	274.3	260.1	3 565.9	3 387.6	
1.6 multiplier	438.9	417.0	5 705.7	5 420.4	Stage II
3.0 multiplier	822.9	781.8	10 697.7	10 162.8	
1.8 seedbed factor		750.6		9 747.8	Stage III
		1 407.2		18 294.1	

Notes: * Half of the year's royalties and licence fees; ** 13 jobs per $1 million expenditure.

FUTURE RESEARCH AGENDA

Like most research undertakings, this one, too, leaves us with perhaps more questions than existed at the beginning. Three areas, especially, suggest themselves for further research: refinement of the seedbed and cluster analysis, improvement of the empirical analysis of clustering and seedbed effect of alliances, and application to policy analysis.

First, the assertion, clearly valid, is made that the very presence of a research university has long-term effects as it creates new knowledge, educates students and trains staff. At the same time, the university attracts specific types of investment, which in turn change inter-industry relationships. In order better to take account of these complex relationships, powerful clustering and seedbed models must be built.

Second, the empirical analysis implicitly attributes all high-tech activity in the region to the presence of the research university and its alliances. In this way their economic effects are likely to be overestimated, since some of the activities and their growth would have occurred anyhow. Adjustments are needed to obtain estimates of net effects. They can be made by assuming that without alliances the region would have enjoyed development typical of the entire state. Another approach would be to look at high-tech industry of a region lacking a research university, such as Portland, Oregon and its high-tech activities.

Third, research into policy applications may be illustrated by an exploration of whether state funding of research at the margin is likely to be cost-effective to state government. Specifically, to what extent will a state allocation, for example, of an additional $100 million to research

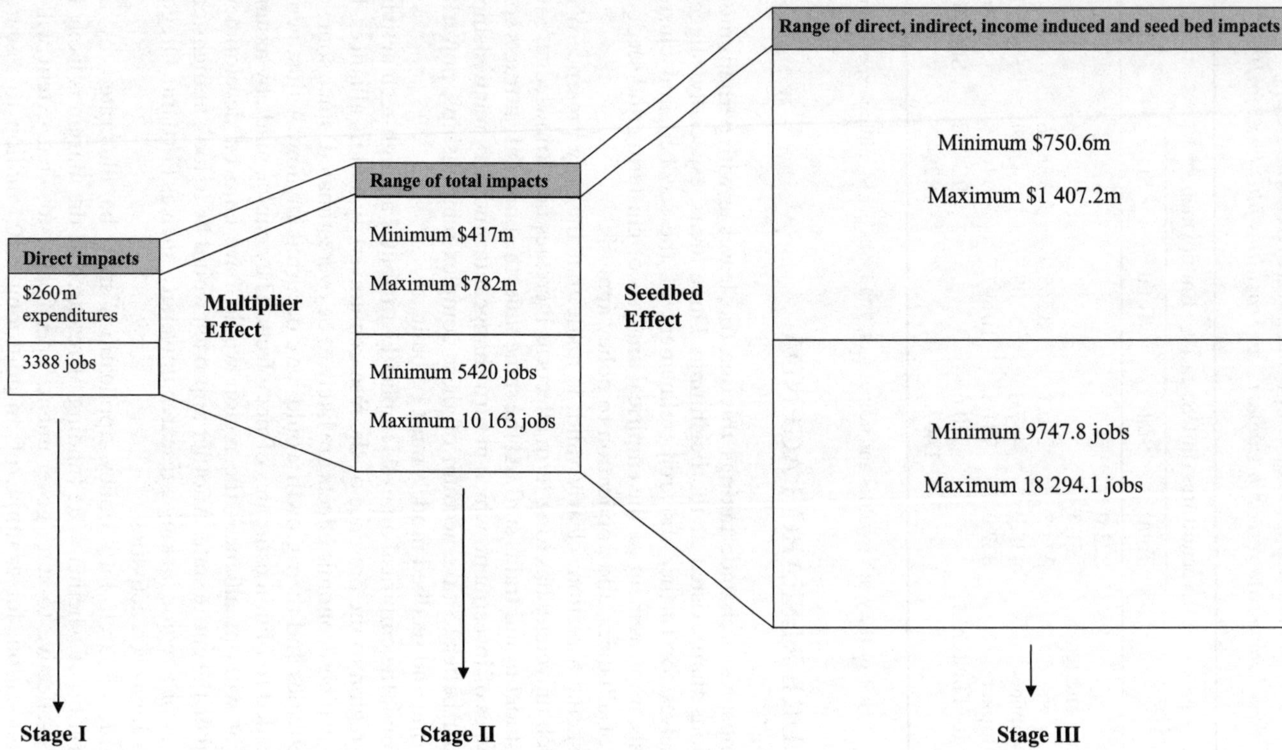

Figure 10.2 Three stages of California's research alliances' economic impact on California

universities further improve their excellence and thus attractiveness to high-tech firms to enter into alliances? Will the additional $100 million in state funds result in state tax receipts growth of more than $100 million and, if so, by how much? From the increased revenue the cost of increased state-funded services must be deducted. The same must be done in relation to the resulting fiscal effects of the state's local governments.

CONCLUSIONS

As the external walls of research universities become more permeable and begin to crumble, and as the academic ivory tower becomes more and more an artifact of the past, great challenges face the university. Likewise great challenges face high-tech industry, which in order to profit must be at the forefront of inventions and innovation. Yet values, history and objectives of the two partners in collaborative undertakings must be harmonized, which will prove to be no mean undertaking. However, if successful, the gains to the partners as well as to regional and national economies can be great. As globalization proceeds, countries who do not partake in these new collaborative undertakings are likely to be left behind in the 21st century. Should that happen, their people's well-being will be put at risk. Thus ways must be found and agreements perfected that serve both well. In other words, both university and industry must be convinced that they can expect net gains. Clearly there is room for government to take steps which facilitate a mutually beneficial outcome. When such conditions exist and research alliances are formed, economic spillovers are likely to result, and so is a clustering of economic activities. The resulting spillovers, which are consistent with those found in the Health Industry Model of Di Tommaso and Schweitzer (2001), have been shown to contribute greatly to the regions' economy and their people's well-being (Bailey and Lawrence, 2001).

NOTE

1. Half the royalties and licence fees are estimated to be related directly and indirectly to corporate research funding in preceding years. While more than half of the university's research funding comes from sources other than corporations (and includes funds for the social sciences, humanities and the arts), work supported by corporations is significantly patent-oriented.

REFERENCES

Bailey, M. and Lawrence, R., 'Do we have a new economy?', *The NBER Digest*, **2**, August 2001.

Business–Higher Education Forum, 'Working together, creating knowledge', **32**, 2001.

Decision Support Systems, 'The economic impacts of the University of California, Irvine on Orange County, 1982–83', December 1983.

Di Tommaso, M.R. and Schweitzer, S.O., 'The health industry: more that just containing costs', background paper presented at the First Ferrara Health Industry Policy Forum, Ferrara (Italy), 21–2 May 2001.

Glover, J.W. and Bibbens, T.E., 'Developing high-technology communities in San Diego', Office of Advocacy Small Business Administration, Washington, DC, January 2000, p. 96.

Harvard Magazine, 'Profiting from the human genome', July–August 2001, 70.

Hirsch, W.Z., 'University high-tech alliances in California: gains and losses', California Policy Options, UCLA School of Public Policy and Social Research. Los Angeles CA, 2000, 45–59.

Marcus, J., 'Universities and private firms cash in on faculty research', The Associated Press, 18 February 1998.

Markoff, J., 'California sets up centers for basic scientific research', *New York Times*, 8 December 2000, A20.

Marshall, A. and Marshall, M.P., *The Economics of Industry*, London: Macmillan, 1885, p. 53.

Marwick, P., 'UCLA Economic Impact', Los Angeles, January 1990, 46.

National Science Foundation/SRS, 'Survey of research and development expenditures at universities and colleges, fiscal year 1997', Washington, DC, 1998, Table B-35.

Shaywitz, D.A. and Ausiello, D.A. 'The necessary risks of medical research', *New York Times*, 29 July 2001.

The Economist, 3 July 1999, 71.

11. Multinational enterprises and high-tech clusters in the health industry: some preliminary results in Italy

Marco Bellandi and Nicoletta Tessieri

1 INTRODUCTION[1]

The success of modern industrial districts, in Italy and elsewhere, represents a clear manifestation of the forces of local development. The industrial districts are prototypical examples of localities (the territorial level is that of daily urban systems) characterized by the economic and social prominence of an industrial cluster of specialized small to medium-sized firms. Within an industrial district, the principal industrial cluster corresponds, in statistical terms, to the aggregation of the most important manufacturing sector of the area and of complementary and auxiliary sectors.

Generally the relations between large firms and the development of industrial districts are various, positive or negative, sometimes relevant. Of course their nature depends also on socioeconomic and institutional factors at regional, national and international levels. As regards the large firms, and specifically multinational companies, there are three different strategies. The first is a strategy of vertical integration in the constitution of human capital, marketing channels, R&D facilities, taking basic resources from the outside and transforming them into specific assets. The second is a networking strategy, in which the internal processes are complemented by the relation with external processes, both in other large firms (joint ventures and so on) and in industrial districts. The last is a predatory strategy, by which external processes are exploited for incorporating valuable external resources and destroying the economic bases of independent districts.

In the first case, localization matters because of the territorial differences in the prices of non-transferable resources, in the local demand for final goods, in the transport costs of transferable resources and products, and in the types and weight of public regulations (classical location problems). In the second case, localization is important also for the reciprocity of progressive relations, in terms of know-how, creativity and market opportunities, with the resources of independent clusters. In the third case, localization is a sort of Trojan horse.

The second case includes what we call here the 'cluster effect'. A condition necessary for the well-being of a district is the presence of a fabric of relations among industrial producers, traders, bankers, policy makers and other subjects, supporting the realization of production projects in teams of specialized companies. The multiplicity of nuclei of specialization acts as an open knowledge laboratory, where new varieties of materials, instruments, organizational solutions and products are continuously experimented with, selected and conserved. The entrepreneur reads the signs of the current changes in the market, collaborating with the people working with him, looking for creative collaboration. The development of new productive knowledge, from the combination of production and trade know-how, learning on the job and technological competencies, is favoured by the same local fabric of economic and social relations and institutions. This way is typical of the industrial district, and corresponds to the deep mobilization of sources of productive knowledge different from (and possibly complementary to) R&D. The embeddedness of the multinational enterprise (MNE) local entities in such local relations may be justified precisely by the advantages of exchanging competencies and ideas in an environment rich in contextual know-how, team capabilities and entrepreneurial spirits.[2]

Outside industrial districts, the role and presence of large firms is normally larger. Clearly this is the case of company towns, or other territorial poles characterized by the social and economic prominence of the local entities of such firms. Mixed conditions are found in other localities characterized by considerable specialization in manufacturing industries (for example, where medium-sized plants and companies have a large role) and especially in large metropolitan areas with their dense tertiary core and diversified industrial structure. In the mixed cases it also possible to find relations and effects similar to those found in the industrial clusters of the districts, together with cluster effects of a different nature.

Within an industrial cluster the productive core is a set of interlinked manufacturing processes and nuclei of contextual know-how. A different case is when the 'productive' core is a set of interlinked technological and scientific knowledge shared by the main actors through a common scientific–technical language. These are 'high-tech clusters'.[3] In the industrial cluster, a central institutional role is usually played by the public and collective organizations that facilitate the exchanges of intermediate products on the local markets and the reproduction of artisan competencies. In a high-tech cluster, a central institutional role is played by universities and other centres of research, culture and high education, which safeguard the public sphere of accumulation and exchange of general scientific knowledge between specialized manufactures, knowledge-intensive services, public services facilities, non-profit organizations and so on.

The two types of cluster are somehow related to a distinction included in the 'Health Industry Model' (Schweitzer and Di Tommaso, this volume), between clusters devoted either to health care products or to health care services. We maintain that the former are nearer to the type of the industrial cluster ('industrial health clusters'), and the latter nearer to the high-tech type ('high-tech health clusters').

This chapter presents an empirical investigation and a descriptive analysis of the presence, in different types of Italian localities, of a set of multinational enterprises (MNEs) dealing directly or indirectly with the manufacturing of health care products. We focus on the possible relations between the MNE local entities (manufacturing plants, research centres and headquarters) and the industries characterizing the localities where they operate. Among the many factors that explain the MNE localization choices, we try to find evidence of the attempt to embed their local entities in health clusters. We also try to find evidence of the presence of the two types of 'cluster effects'; that is, localization justified by the specific advantages of contact either within an industrial cluster or within a high-tech cluster, even if it is not easy to identify and distinguish them in empirical terms.

The following section illustrates the methodological framework of the empirical investigation. Section 3 presents the definition of the sectors under examination (the 'health industry'), of the main multinational groups with their local entities and of the types of localities (local systems). Sections 4 to 6 consider the data on the localization of MNE entities in Italian regions and localities, first identifying some general characteristics, then the relation between those entities and the sector specialization of the local systems, and finally the relation between the same MNE entities and local high-tech characteristics. Section 7 presents an evaluation of the signs of either industrial cluster effects or high-tech cluster effects. Sections 8 and 9 include a summary of general results and some concluding remarks on perspectives of research and industrial policies for the health industry. Box 11.1 in the appendix presents a summary of and keys to the various acronyms used in this chapter. Tables 11.1 to 11.4 present some of the data discussed in sections 4 to 7.

2 THE METHODOLOGICAL FRAMEWORK

The empirical investigation looks at the location choices in different types of Italian local systems of the main multinational enterprises, whose activities are directly or indirectly connected to health care products or to health care services. At the same time we try to find signs of the embeddedness of their local entities in health clusters as indicated in the previous section.

This set of MNEs is divided into four sub-sets: dedicated (specialized in the health industry), non-dedicated (diversified, but including activities in the health industry), only potentially connected (possible indirect connections with the health industry) and definitively non-connected.

A first set of local systems (1LS) is identified by the presence of local entities (manufacturing plants, headquarters, research units) of companies of 'dedicated' multinationals, and/or by the presence of companies (that is, their local entities) working in the health industry but belonging to 'non-dedicated' multinationals. These two types of companies are indicated by the acronym HIB ('health industry's branches').[4]

The simple interpretative hypotheses on which we base our empirical test are the following.

1. We assume as proxies of some 'industrial health cluster effect' cases where there exists a sectoral relation between the activities of the MNE local entities and the industries characterizing the economic structure of the local systems with a district like industrial structure (manufacturing specialization, prevalence of small–medium firms).
2. In other types of local systems not dominated by large plants and firms, such sectoral relation is considered not enough to suggest the presence of some industrial cluster effect. However, the evidence is strengthened if, in the same local systems, both headquarters and plants are found. When we find plants only, this is assumed to be a sign of a low degree of embeddedness.
3. Independently of the nature of the firms and their strategy (predatory or not predatory), the propensity to establish relations with the context is higher when more headquarters (with different activities too) are present at the same time.
4. Production and innovation linkages within and between sectors may be supported in the absence of the above correlation, if the local system is characterized by high-tech activities, even outside the health industry, being the expression of an environment rich in knowledge, R&D, telecommunication infrastructures and various types of high-tech service activities. So we assume as a proxy of the presence of a 'high-tech cluster effect' local systems characterized by these types of sectors and where more than an HIB headquarters is observed.

In order to check the diffusion and relevance of the cluster effects, the indexes of specialization in the pharmaceutical and biomedical sectors shown by the local systems are controlled.[5] In particular, we look at the intersection between 1LS and the local systems where the index of specialization in the pharmaceutical or/and biomedical industry is higher (here the threshold is set arbitrarily at 2.00). Furthermore we also look for the 1LS local systems characterized by high-tech activities among the first three sectors of specialization.[6] The presence of headquarters and plants is also analysed and commented upon.

3 THE UNITS OF INVESTIGATION

This empirical investigation is based on three units of analysis: the Italian local systems, the main Italian and foreign multinational enterprises and the health industry. The latter is an aggregate of sectors including not only the pharmaceutical and biomedical ones, but also other manufacturing activities that may be connected to the first ones.[7] 'Local systems' are defined as localities showing a high degree of social and economic interaction between the production sites and the residential sites.[8] We adopt here the set of local systems identified by Istat-Sforzi with data on commuting flows in 1991.[9]

Different spatial configurations, different productive and socioinstitutional structures, different degrees of connection between the local community and the industrial clusters identify different types of local systems hosting different types of clusters.[10] Here we restrict the analysis, adopting a classification of the Italian local systems (identified with data from Istat, 1991) based on data from Istat, 1996, 'Intermediate Census'. These data essentially look at the type of sector specialization and the size of the units characterizing the local productive structure.[11]

According to this classification, the local systems are distinguished in manufacturing systems (MLS) and non-manufacturing ones (NMLS), depending on the presence of manufacturing specialization, from the national distribution of employment in manufacturing and tertiary (non-public) sectors. Among the former, local systems of small–medium (SMLS), medium–large (MMLS) and large (LMLS) enterprises[12] have been distinguished with the help of a database developed by Giovanni Solinas.[13] Among the NMLS, we have distinguished the large metropolitan areas (MELS)[14] and other non-manufacturing local systems (ONMLS).

As far as the manufacturing multinational groups are concerned, we define them as webs with a parent company also controlling manufacturing subsidiaries in different countries. We have used the database published in an annual directory by *Ricerche e Studi* (R&S), and covering the 182 most important groups headquartered in Italy.[15] R&S data for two sets of MNE groups have been extracted: (a) those controlled by Italian shareholders with multinational manufacturing or research sites, and (b) those controlled by foreign groups with companies' headquarters, plants or research sites in Italy.[16]

4 SOME GENERAL CHARACTERISTICS OF THE HIBS AND THEIR PARENT GROUPS IN ITALIAN LOCAL SYSTEMS

In 2000, out of the 182 'important' groups registered in the R&S database, 18 multinational groups are 'dedicated' to the health industry, or 'non-dedicated' but with companies connected to the health industry.[17] Within these groups, the companies connected to the health industry (HIBs) number 91, with 85 headquarters and 72 manufacturing plants, in 50 local systems (the ILS), that is, 6.4 per cent of the total number of the local systems.[18] Within the 18 groups, seven are under foreign control.[19]

The foreign ownership does not seem to imply a peculiar influence on the location choices. HIBs of 'foreign multinationals' are found in 33 out of 50 local systems (66 per cent); HIBs of 'Italian multinationals' in 29 out of 50 local systems (58 per cent). Italian and foreign multinational groups are jointly present in 18 local systems.[20] The preference for the great metropolitan areas, and more generally for local non-manufacturing systems, is confirmed in both types of HIBs.[21]

At a regional level, the higher number of local entities, either headquarters or manufacturing plants, belonging to HIBs is found in Lombardy, then in Tuscany, Emilia-Romagna, Latium, Abruzzo and Veneto (see Table 11.1). The ratio of manufacturing plants to headquarters is different from region to region. This seems also to be associated with the greater or lesser presence of non-manufacturing local systems, and in particular of metropolitan areas.[22]

As regards the degree of concentration of multinationals within the local systems, only 18 out of 91 HIBs (19.8 per cent) have both headquarters and manufacturing plants in the same locality. This happens in 13 local systems out of 50, either manufacturing or non-manufacturing. This rich presence of business functions is here interpreted as an indicator of a degree of territorial embeddedness of the multinational companies.

Only nine out of 18 multinational groups have more than one company in the same local system. This happens in seven local systems. Furthermore, 14 out of 91 companies (15.4 per cent) belonging to 11 out of 18 multinational groups (61 per cent) operate through manufacturing plants in more than one local system. The multi-localization could generate cross-fertilization between local systems of various types. Only one of the

Table 11.1 HIBs by regions, provinces and local systems

Region	Province	Local system	MNEs	HIBs	Headq	Plants
Piemonte	Alessandria	Alessandria	1	1	0	1
Piemonte	Alessandria	Casale Monferrato	1	1	1	0
Piemonte	Cuneo	Ceva	1	1	0	1
Piemonte	Vercelli	Crescentino	1	1	0	2
Piemonte	Torino	Torino	3	3	1	3
Lombardia	Mantova	Mantova	1	2	1	2
Lombardia	Pavia	Pavia	1	1	0	1
Lombardia	Bergamo	Albino	1	1	0	1
Lombardia	Bergamo	Bergamo	1	1	1	1
Lombardia	Varese	Busto Arsizio	1	1	0	1
Lombardia	Como	Como	1	2	2	3
Lombardia	Cremona	Crema	1	2	2	1
Lombardia	Milano	Desio	2	2	0	2
Lombardia	Como	Lecco	1	1	1	1
Lombardia	Milano	Milano	9	32	31	5
Lombardia	Pavia	Voghera	1	1	0	1
Liguria	Genova	Genova	1	2	1	1
Veneto	Rovigo	Castelmassa	1	1	0	1
Veneto	Padova	Padova	1	1	0	1
Veneto	Verona	Verona	3	6	5	2
Friuli.Ven. Giulia	Pordenone	Pordenone	1	1	0	1
Emilia-Romagna	Bologna	Bologna	2	3	2	2
Emilia-Romagna	Parma	Fornovo di Taro	1	1	0	1
Emilia-Romagna	Modena	Mirandola	1	2	2	2
Emilia-Romagna	Parma	Parma	4	4	1	5
Emilia-Romagna	Reggio Emilia	Reggio Emilia	1	1	1	0
Marche	Ancona	Ancona	1	1	0	1
Marche	Ascoli Piceno	Ascoli Piceno	1	1	0	1
Marche	Ancona	Jesi	2	1	0	1
Toscana	Arezzo	Arezzo	1	1	0	1
Toscana	Firenze	Firenze	2	13	12	3
Toscana	Pisa	Pisa	1	2	2	0
Toscana	Siena	Siena	1	1	0	1
Lazio	Latina	Aprilia	1	1	0	1

Table 11.1 (continued)

Region	Province	Local system	MNEs	HIBs	Headq	Plants
Lazio	Frosinone	Frosinone	2	2	0	2
Lazio	Latina	Latina	1	1	1	1
Lazio	Roma	Roma	4	9	7	2
Abruzzo	L'Aquila	L'Aquila	2	2	1	1
Abruzzo	Pescara	Pescara	3	4	4	3
Abruzzo	Chieti	Vasto	1	0	0	1
Campania	Caserta	Aversa	1	0	0	1
Campania	Caserta	Caserta	1	2	1	1
Campania	Napoli	Napoli	1	1	1	1
Puglia	Bari	Bari	1	1	0	1
Puglia	Brindisi	Brindisi	2	2	1	2
Sardegna	Cagliari	Cagliari	1	1	0	1
Sardegna	Cagliari	Villacidro	1	1	0	1
Sicilia	Catania	Catania	1	1	1	2
Sicilia	Trapani	Marsala	1	1	1	0
Sicilia	Palermo	Palermo	1	2	1	1
Italy total		50			85	72

18 multinational groups has more than one company with headquarters carrying out mainly research, and this happens within a single large metropolitan area (Rome). Generally headquarters carrying out research activity are located in foreign countries.

As far as the size of the manufacturing plants of the HIBs is concerned, the prevailing size is medium and small, which is similar to that generally characterizing the production structure of local manufacturing systems where the majority of them are located.[23] Thus, the presence of these multinational companies does not tend to modify the size structure of the manufacturing industries of the local systems;[24] they seem often to 'mimic' within the local context.[25]

5 THE RELATIONS BETWEEN HIBS AND THE LOCAL SPECIALIZATION IN THE HEALTH INDUSTRY

We consider now the possible presence of some correlation between HIBs (and their parent groups) and the sectoral characteristics of the manufacturing industry of the local systems in which they operate.[26]

First of all, let us mention that the sector with the highest frequency, within the first three sectors of local manufacturing specialization of the 50 1LS local systems, is the pharmaceutical (and chemical products for medical uses) sector (244 in Ateco Istat). In particular, 31 of the 50 1LS (62 per cent) show a specialization index superior to one for the pharmaceutical or biomedical sector. Only 13 out of 50 1LS (26 per cent) have an index of specialization superior to unity for both sectors.[27]

Out of the 18 MNEs more closely connected to the health industry,[28] 10 have headquarters (of controlled companies) in systems which have the pharmaceutical sector among the first two manufacturing specialized sectors;[29] four out of these 10 groups are 'dedicated groups'. Furthermore six of these 18 MNEs have manufacturing plants in local systems that have the pharmaceutical sector among the first two manufacturing specialized sectors.[30] Only one group appears to have headquarters and manufacturing plants in a local system that has the biomedical sector as the first sector of manufacturing specialization.[31]

In order to check the diffusion and relevance of industrial cluster effects, suggested by the data just recalled, we consider all the local systems where the index of specialization in the pharmaceutical or/and in the biomedical sectors is particularly high (here the threshold is set arbitrarily at 2.00). They number 80. We denote them simply as 'local systems specializing in the health industry'. Let us now consider some evidence with the help of various sub-sets. The first three are a complete partition of that set of 80 local systems; the fourth sub-set represents another partition of the same set, the complement not being recalled here (Tables 11.2, 11.3 and 11.4).

Table 11.2 Types of local systems involved by the presence of HIBs

Types of local system	Number of local systems involved by the presence of HIBs	Total number of local systems, by typology	% of local systems involved by HIBs
SMLS	11	193	5.70
MMLS	14	78	17.95
LMLS	2	21	9.52
Total MLS	27	292	9.25
ONMLS	13	481	2.70
MELS	10	11	90.91
Total NMLS	23	492	4.67
Total LS	50	784	6.38

Table 11.3 Headquarters and manufacturing plants belonging to HIBs, by types of local systems

Type of local systems	Number of headquarters	Number of plants
LMLS	2	5
MMLS	6	19
SMLS	6	14
Total MLS	14	38
ONMLS	15	15
MELS	56	19
Total NMLS	71	34

Table 11.4 Local systems (LS) belonging to 1LS, by types, with pharmaceutical (PSI) and/or biomedical specialization (BSI) indexes higher than 2.0

Local system	LS	PSI	BSI
CRESCENTINO	MMLS	0.899	7.571
CEVA	MMLS	6.562	0.278
GENOVA	MELS	0.485	2.086
MILANO	MELS	5.535	1.755
VERONA	ONMLS	2.553	1.089
PARMA	MMLS	2.194	1.701
MIRANDOLA	SMLS	0.566	13.973
BOLOGNA	MELS	0.222	2.516
ANCONA	ONMLS	4.426	1.708
FIRENZE	MELS	2.740	1.462
PISA	ONMLS	10.384	1.964
SIENA	ONMLS	13.294	1.207
ROMA	MELS	4.570	2.831
APRILIA	MMLS	17.561	0.858
LATINA	MMLS	9.394	1.860
FROSINONE	MMLS	3.439	0.627
L'AQUILA	ONMLS	13.228	0.880
CATANIA	MELS	5.064	1.885
CAGLIARI	MELS	0.022	2.065

1. *The intersection with 1LS* (19 local systems). Out of these 19 local systems, 12 are non-manufacturing (seven metropolitan).[32] Among the seven local manufacturing systems, six are local systems of medium enterprise and one is a local system of small enterprise. The latter,

Mirandola, registers the presence of both HIB headquarters and HIB plants. Among the MMLS, only Parma and Latina have both a headquarters and plants; Aprilia, Ceva, Crescentino and Frosinone have plants only. The headquarters of MNEs that operate for the health industry are more numerous in local non-manufacturing systems. The types of local systems more frequently showing a specialization index higher than 2.0 (in the pharmaceutical and/or biomedical sector) are (in absolute terms)[33] the local non-manufacturing systems, especially the metropolitan ones. A considerable number of HIB headquarters are concentrated in three MELS (Milan, Florence, Rome) and, though on a smaller scale, in one ONMLS (Verona) included in this sub-set: 55 out of 85 (for all local systems). In the same local systems there are only 12 out of 72 HIB plants (for all local systems).

2. *2LS* (5 local systems). Outside 1LS, four non-dedicated multinational groups have companies with only potential indirect connections located in five of the 80 local systems specializing in the health industry; that is, the 2LS set.[34] Here three are non-manufacturing local systems, but none is in the great metropolitan areas. Of the two local manufacturing systems, one is a local system of large enterprise and one is a local system of medium enterprise. All but one of the local entities are manufacturing plants; furthermore only one group operates through a manufacturing plant in a local system where the pharmaceutical sector is among the main specialized sectors.[35] We have confirmation that the presence of only potential and indirect connections with the health industry among the activities of a MNE group produces a poor fit with the presence of local specialization in the health industry.
3. *Local systems specializing in the health industry not included in 1LS or 2LS* (56). The other 56 local systems[36] are mainly non-manufacturing ones. Only seven are local manufacturing systems; five are local systems of small enterprise and two of medium enterprise. As far as the indexes of specialization are concerned, seven local systems have a very high index of specialization in the pharmaceutical sector, four have a very high index of specialization in the biomedical sector. These data confirm that the presence of local specialization in the health industry (biomedical and pharmaceutical sectors) is largely associated with an aggregate non-manufacturing specialization, even in the absence of local entities of HIBs.
4. *The local systems with the highest specialization in the health industry* (26).[37] The highest sectoral specialization of the local systems in the health industry does not necessarily go with the numerous presence of local entities of the multinationals' companies (HIBs) directly or

indirectly connected with the health industry – not even in cases of scientific/academic milieux such as Pisa or Siena.[38] Few are the HIBs operating in local systems where the indexes of specialization in pharmaceutical or biomedical products are higher than 9.0.[39] However, HIBs are present in 15 of these local systems.

6 THE RELATION BETWEEN HIBS AND THE LOCAL SPECIALIZATION IN HIGH-TECH INDUSTRIES

Sector similarities give an important clue to the existence of linkages at the production and market levels. They may be relevant also for innovation games and outside health industrial clusters too.[40]

However, the health industries, as they have been defined here, are *mainly* intensive in high-tech activities. This could explain the location choices of MNE within the diversified structure of the metropolitan areas, where knowledge-intensive business services often cluster. So we try to explore the existence of a possible correlation between HIBs and the relative specialization of local systems in high-tech sectors. The hypothesis is that a local system rich in high-tech activities has good infrastructure for research and knowledge exchanges. Furthermore the existence of various high-tech sectors may be a source of innovative local relations, especially as far as firms producing novel products are concerned.

In order to explore these relations, the first three sectors of specialization (see note 6) of the 50 1LS local systems have been reclassified within four macro sectors using a Pavitt-like typology (Pavitt, 1984).[41] Some relations seem to emerge from a simple descriptive analysis.

Considering the set of the first three sectors for all 1LS, the Pavitt sector most frequently represented is traditional industry, followed by high intensity of scale, high-tech and, last, specialized suppliers. Considering only the first sector of manufacturing specialization gives a different order: high-tech, traditional industry, high intensity of scale and specialized suppliers.

Let us consider the main agglomeration of HIBs. In the local systems of Milan and Rome, where the highest number of multinational groups is located, the first two Pavitt sectors are high intensity of scale and high-tech; but in Milan, within the three sectors of specialization, two are high-tech and in Rome, two are high intensity of scale. Similarly, in Turin, we find high intensity of scale and high-tech. The first three sectors of specialization of the local systems of Verona are in the high intensity of scale and specialized suppliers classes. The first three sectors of specialization of the local system of Florence are the traditional industry and the high-tech one; in Pescara, we have high intensity of scale and traditional industry; only in Parma are all three sectors of specialization in a single class (traditional industry).

Finally we have some evidence of an *independent* role as an explanatory variable of HIBs' location, of a local environment rich in high-tech activities. In any case, the results suggest the necessity to investigate in more detail the 'high-tech' characteristics of the local systems where HIBs are particularly concentrated. Some clues can be extracted by comparing the 1LS with a map of large geographical concentrations of patented innovation sources in Italy.

The map comes from research conducted in 1995. It was aimed at the identification of large geographical areas showing signs of a high internal dynamism in innovation (and exports) in connected sets of manufacturing and tertiary sectors.[42] We consider only the areas associated to sectors directly or indirectly related to the health industry. The map of the 'innovative systems in the pharmaceutical sector' shows, in a dominating position, the system centred on the province of Milan and extending, among the other provinces, to Como and Pavia. The local systems of Milan, Como and Pavia also appear, according to our analysis, to host HIBs. In particular we know that Milan is the main agglomeration of HIBs in Italy, and we should recall here the role played by various public organizations, public and private research centres, among them the CNR, the San Raffaele Centre, the Cise and the Carlo Besta Foundation. In second position there is the 'the innovative system' which has the province of Rome at its centre. Rome, after Milan, and with Florence, hosts lots of MNEs, HIBS and HIBS headquarters.[43] A third innovative system connected to the pharmaceutical sector is in Triveneto, including, among others, the provinces of Verona and Padua. Still in the pharmaceutical sector, another 'innovative system' is the Tusco-Emilian where, among others, we find the provinces of Florence, Siena and Bologna. In the local systems centred in the cities just cited there are several HIBS.

For the 'innovative systems of biomedical machines', a wide system including the province of Pordenone (where the Bormioli Group operates) and Veneto, Emilia Romagna and Tuscany provinces is defined. Among the other four innovative systems of the biomedical sector, the most important one is that centred on Milan, followed by the 'Rome system', comprising the provinces of Rome and Latina, the system of Turin including the provinces of Turin, Vercelli and Cuneo, and lastly the Sicilian area, with the provinces of Catania and Palermo.

A last reflection is suggested by intersectoral overlapping, especially with other high-technology sectors. It appears particularly strong in the 'Milan system' again.[44]

Thus, from the comparison of the map of innovative large geographical sources of patented innovation connected to the health industry, and the map of local systems hosting the MNEs more closely connected to the health industry, there appears a correlation between the MNE location choices and the existence of an innovative and dynamic local environment also oriented towards the health industry.

7 CLUSTER EFFECTS

Let us take again the main empirical points defined in the previous sections. On the one side, there are several signs of the presence of a slight relationship between the MNE location choices and the specialized manufacturing sectors of local systems. This relationship could be the image of location strategies aimed at taking advantage of the contact with the *industrial health cluster effect.*

1. The highest frequency, within the set of the first three sectors of local manufacturing specialization in 1LS, is found in sectors of the pharmaceutical and chemical products for medical uses.
2. Within the 1LS set, the presence of HIBs is higher in non-manufacturing local systems than in manufacturing local systems. This could be correlated to the fact that in the first types of local systems the indexes of specialization in the pharmaceutical sector and/or in the biomedical one appear to be superior to 2.0 in a proportion higher than in manufacturing local systems. Moreover, in the first type of local systems, the tertiary activities dominate and the advanced tertiary sector is particularly concentrated.
3. However, even in a few manufacturing systems, there are signs of an industrial cluster effect. This is suggested both by the high indexes of specialization in the health industry and by the simultaneous presence of HIB headquarters and plants related to the same MNEs.[45]

As we have seen, other evidence goes towards the relevance of the *high-tech cluster effect.*

4. The manufacturing local systems of 1LS have the 'traditional' sectors (Pavitt classification) in the first rank, and the 'high intensity of scale' sectors in the second.[46] Instead, among the non-manufacturing local systems, the 'high intensity of scale' sectors are first and the 'high-tech' sectors second.
5. The headquarters of HIBs seem relatively more concentrated in non-manufacturing areas than in the case of other multinationals.[47] The opposite relations prevail for manufacturing plants.

6. The location of the headquarters is most prominent in the great metropolitan areas (Milan and Rome, but also Florence). This is easily related to the local presence of sophisticated customers, such as large hospitals and sanitary centres,[48] sources of knowledge related to the presence of university, research and telecommunication centres and sources of financial resources related to the local presence of big banks. More generally, it is related to the presence of a business environment, rich in services and transportation infrastructures, which constitutes a node for combining resources and knowledge from different actors and contexts, regional, national and international. This is confirmed especially by observing that most of the headquarters of MNEs belonging to any sector or industry are located in Milan.[49]
7. Comparing the 1LS with the map of 'innovative large geographical sources of patented innovation' connected with the health industry, there is evident a correlation between the location choices of HIBs and the existence of innovative and dynamic local environments also oriented towards the health industry.

The descriptive analysis reported in this chapter (on the relations between the different types of local systems and the location choices of big MNEs connected to the health products and health care services) has several limitations. We could not distinguish between greenfield or cross-border investments; we had not the support of systematic information on how long the local entities of MNEs (with an endogenous or exogenous origin) have been located in these local systems, or on how great is the effect of multinationals on local employment (and turnover, exports and so on); and we have not considered the internal organization of the MNEs involved, or the influence of the HIB-specific sectoral characteristics on their location and embeddedness strategies.[50]

All this also means that we could not distinguish whether the cases of MNE local entities signalling linkages at the local level are driven either by a strategy of embeddedness or by other types of strategies. However, even if deeper empirical investigations and more structured statistical analyses may be necessary, some useful elements about the types of strategies adopted by these MNEs have emerged. In particular:

8. Most of the 18 big MNEs connected with the Italian health industry do not carry out research activities in Italy. On the other hand, eight of these MNEs have located more than one company inside the same local systems. This suggests a somewhat rich set of linkages.
9. There are 22 (out of 50) local systems where HIBs are located that seem to be solely the object of production decentralization strategies (only

manufacturing plants), probably aimed at taking advantage of cheap labour, cheap land or public incentives. This is normally outside a virtuous circle of cluster effects and local development.

10. Among the cases of local systems hosting only HIB manufacturing plants, there are four MMLS, one MELS and two ONMLS with a high index of specialization in the pharmaceutical sector or in the biomedical sector.[51] This may be a sign of two alternative courses of action. The first could be looking for traditional incentives or low cost of factors (within a strategy either of the vertical integration type or of the predatory type). The second could be after specific competencies and infrastructures within local health industrial clusters (perhaps still within a strategy of the predatory type,[52] but not necessarily).

Summing up, specialized sectors and types of local systems would appear to be clearly correlated to the localization choices of the multinationals' companies connected to the health industry (HIBs). *The sector specialization at the local level is particularly important, as a condition of location of headquarters, or headquarters together with manufacturing plants (signalling stronger local linkages), when it is related to other conditions of local environment supporting knowledge-intensive exchanges in fields related to the health industry*. The preference towards local environments in which the specialization in the health industry is high but not the highest, combining with other high-tech and non-high-tech manufacturing sectors, and with the manifestation of innovative capacity and strong activities in health services, confirms the idea. More precisely, it suggests the presence of local entities of HIBs that are looking for high-tech cluster effects, perhaps combined with health industrial cluster effects.

Clear signs of pure health industrial cluster effects concern a very limited number of local systems.[53] To these cases we have to add those where weaker linkages at the local level (viz. only manufacturing plants) show, nonetheless, a connection with a high local specialization in the biomedical or pharmaceutical sector.

8 HINTS FOR RESEARCH PERSPECTIVES AND POLICY IMPLICATIONS

Some hints for future studies and consequences in terms of industrial and local development policies are now illustrated briefly. At the MNEs level, further research is necessary on the influence (on the location choices and on the embeddedness strategy) played by the different types of firm/group (structure/strategy/performance) and the different types of sector/industry where HIBs and MNEs operate. Connected with this is the research on the nature, degree and stability of the linkages of the MNEs with internal and external, local and extralocal stakeholders. They could generate cross-fertilization between different contexts.[54]

At the territorial/sectoral level, research is necessary on the role that different types of public good and services, and formal and informal institutions, could play in attracting MNEs and stimulating their embeddedness in a local environment, rich with conditions neither easily reproducible/tradable nor easily internalized by just a few economic and social actors. In such an environment, profitable exchanges and combinations of resources, competencies and knowledge, can involve MNEs and local and extralocal stakeholders.[55] The possibility of a virtuous circle of embeddedness in cases of stronger linkages should be supported by stimulating the cross-fertilization between different types of environment and different types (in kind and size) of enterprises and public organizations, especially with respect to the strengthening of learning and innovation networks.[56] The upgrading of low-quality linkages (only production decentralization) towards a higher knowledge and institutional intensity may be the object of policies aimed at strengthening rich local factors and monitoring and selecting out opportunistic private strategies (Bailey et al., 1999).

At the cluster level, policies should be calibrated to the specific type of cluster, and managed at different territorial levels (local system, region and country level).[57] If we consider, simply, the distinction between a high-tech health cluster and an industrial health cluster, we can focus on a few general aspects. In the first case, an important object is the improvement of cooperation and communication among firms, regulatory bodies, consultancy and technical providers, hospitals, universities, public and private research centres/labs and other organizations of higher education and research. Italian universities, compared to the universities of other countries, usually show a moderate inclination towards direct involvement in business activities (Owen-Smith et al., 2001), and weak support for applied research. However, particular cases may show more promising conditions, and they should be identified in a systematic way, in order to involve them in broader networks of excellence, together with actors and structures in different economic, social and political environments. In the second case, a particular role could be played by collective organizations (local, regional and international), supplying public goods and services (included training and financial support) that can sustain and develop the entrepreneurial capabilities connected with a good quality manufacturing tradition. At the same time, they can support the growth of creative/specialized professional figures, and the appreciation of artisan competencies, making them available within flexible production and innovation networks, also independent of large firms.

9 CONCLUSIONS

At the end of a very detailed chapter, and after a couple of sections defining general results and perspectives of research on positive and normative matters, it is useful to conclude quickly. In fact we would like to propose just two speculative remarks.

Firstly, it is perhaps a case of coming back to the question of the relationship between health industry and local systems. The definition of 'local system' here adopted follows a particular methodology. It may raise doubts regarding its specific results. However, the core of the game is to capture, by approximation, the life of places shaped by daily, material, face-to-face social and economic relations. We have found signs that this game makes sense in some cases, even when big MNEs are on the stage and play as important actors. They are cases of great interest since they seem to correspond to the possibility of highly dynamic conditions, in which the local environment is not the passive host of global corporate strategies. We know that the passive status is largely and surely present in many cases; however, the cases where local capabilities are important show an alternative which helps the reflection on different health industrial policies and different paths of development.

Secondly, the Italian case deserves a qualification. Italy has little specialization in health industries and generally in the high-tech industries and clusters. It is well known that Italian industry has its points of excellence within the so-called traditional industries and in some related mechanical ('specialized suppliers') industries. This has often been seen as a terrible weakness, in academic and government quarters. However, many attempts to support high-tech or scale-intensive industries have had not very impressive results. Below this general representation, there is a reality in which some points of excellence seem to exist also in high-tech and health industry. Traditional and mechanical industries, largely based in industrial districts, are also evolving. They absorb higher degrees of new technologies, opening new routes for interaction with emerging high-tech clusters.

NOTES

1. Marco Bellandi is the author of sections 1 and 9; Nicoletta Tessieri is the author of sections 2 to 8. We gratefully acknowledge the support and advice of the editors and the referees, and the help given in the empirical research by Fulvio Coltorti, Antonio Benzoni, Stefano Menghinello, Giovanni Solinas, Fabio Sforzi, Lorenzo Zanni, and Luciana Lazzeretti. Previous versions of parts of this paper have been presented by N. Tessieri at The II and III Ferrara Forum on Health Industry Policy, November 2001 and 2002, Ferrara, Italy; and at the Eunip Conference, December 2002, Abo Akademi, Turku, Finland. The empirical research here reported, concluded in 2003, is just the first

step of a larger project by the authors, comprising not only statistical research and identification, but also application to case studies and policy implications. The usual disclaimer applies.

2. On industrial districts and local development, see Becattini (2000), Becattini et al. (2003); on multinational (or large) companies and regional development, see Dicken et al. (1994), Vaccà (1995), Porter (1998), Vázquez-Barquero (1999), Enright (2000), Hood and Young (2000), Zanfei (2000), Bellandi (2001), Tessieri (2000, 2003).
3. On high-tech clusters, cf. Benichou and Viens (below), Antonelli (2000), Keeble and Wilkinson (2000), Cooke (2002).
4. Some comments also concern the MNEs carrying out activities apparently not connected with the health industry, but operating (through local entities) within the first set of local systems.
5. The index of specialization for manufacturing sectors is calculated in the following way: $S = (n_{kl}/n_l)/n_k/n)$; where n is the number of employees, k is the sector and l the local system; n_k is national employment in sector k; n is national employment in manufacturing industry. We control also, within those local systems, the ones not included in 1LS and with the presence of multinationals' companies only potentially connected with the health industry. This particular set of local system is indicated, for short, as 2LS. Data on local specialization in services sectors could not be considered at the time we realized this statistical elaboration.
6. We consider the three sectors (at the three-digit level of ATECO Istat, that is, the classification of economic activities used by the Italian national institute of statistics (1991)) with the highest specialization index against the national average in the manufacturing industry. The inner order of the three sectors (I, II and III) is defined by weighting the index of specialization with the quota of employment in the sector on the whole manufacturing employment of the local system. The analysis should be extended to the set of all the sectors, but we use this limitation to look at the evidence of strong signs of correlation. Different results, in terms of possible connections, appear by ordering all the sectors of each local system. This methodology has been applied in other empirical research in order to find signs of the degree and nature of the embeddedness of these MNEs and their role inside each local system: see Tessieri (2003).
7. The first sectors are identified by ATECO Istat 1991: (244) manufacturing of pharmaceutical and chemical products for medical uses; (331) manufacture of medical, surgical and orthopedic appliances. The other sectors correspond, for example, to the manufacturing of rubber articles and plastic materials, such as polyester film used in radiographic products, the manufacturing of non-metallic minerals for glass tubes and bottles, food and drink manufacturing for dietetic products, the manufacture of machines and mechanical appliances (robots, machines for the production of plastic caps, nappies and so on), production of special papers, enamels, production of mini steel products (needles, forceps and so on), production of microelectronics, information technology services (telemedicine) and so on. Considering the 'Health Industry Model', the set of all these sectors corresponds largely, if not completely, to the 'manufacturing of health care products' and to the production of intermediate products, instruments and services for the first ones.
8. See Becattini (2000). A review of the Italian methodologies for the empirical identification of local systems and industrial districts is in Sforzi and Lorenzini (2001). For an application to the UK, see De Propris (2001).
9. Their spatial identification has been carried out on the basis of a technique which selects the smallest areas compatible with the inclusion in each area of a high percentage of house-to-work commuting flows of the population residing in the same area (Sforzi and Lorenzini, 2001). The local systems taken into account here are defined on the basis of the 1991 Census data: it is a set of 784 areas (each corresponding to the territory of a contiguous set of a few Italian municipalities) (Istat, 1997).
10. For details on different types of environments that are preferred by MNEs, depending on the type of their operational units and their main activities, see Tessieri (2003).
11. In order to identify the social and institutional factors characterizing the different types of local systems as well as the degree of connectivity between different actors within and

outside the local systems, a broader set of data and the help of case studies would be needed.

12. The first type is characterized, against the national average, by the relative prevalence of local units of the manufacturing industry with fewer than 100 employees; in the second type, the characterizing size of manufacturing local units is 100 to 499 employees; and in the third type, more than 499 employees.
13. Who has kindly supplied the data for the elaboration presented here. For a general presentation of this database see Solinas and Baroni (2001).
14. As far as the identification of metropolitan systems is concerned, we have assumed the Istat, 1991 classification (Istat, 1997), with the only exception being the local system of Turin that we considered an LMLS, according to the Solinas classification. So far MELS include Milan Venice, Bologna, Florence, Rome, Naples, Bari, Palermo, Catania and Cagliari.
15. *Ricerche e Studi* (R&S) is a Mediobanca subsidiary specializing in economic and financial research. Here we are referring to the 2001 edition of the R&S directory (including figures for the 2000 financial year), which covers 182 groups including the main private sector companies, the principal state-controlled enterprises and the leading multinationals with substantial manufacturing operations in Italy. These groups are selected by the R&S Advisory Committee on the basis of an overall assessment that takes account of size, positioning in the industry concerned, performance (in terms of profits, exports and growth), importance in the capital market and corporate organization.
16. On the basis of the R&S database, the companies controlled through holding 50 per cent or more of shares on capital have been included as companies of the multinational group, whereas the minority shareholdings are excluded, except for some cases worthy of particular attention.
17. The 'dedicated' groups are five in number: A. Menarini Industrie Farmaceutice Riunite, Aventis Pharma, Bracco, Pharmacia & UP John, and Glaxo Wellcome Finanziaria. The 'non-dedicated' groups are number 13: 3M Italia Finanziaria, Air Liquide Italia, Artsana, Bormioli Rocco & Figlio, Cofide, Compart, CSII Industrie, Finaf, Heinz Italia, Procter & Gamble, Saig, Snia and Star.
18. The local systems hosting the higher number of dedicated and non-dedicated groups are, in order, Milan, Rome and Parma; Turin, Verona and Pescara. The highest number of HIBs is present, in order, in Milan, Florence, Rome, Verona, Parma, Pescara and Turin. The highest number of headquarters is found in Milan, Florence, Rome, Verona and Pescara; then Bologna, Frosinone, Brindisi, Desio and L'Aquila. The highest number of manufacturing plants is registered in Milan and Parma; a lower number is registered in Turin, Florence and Pescara. With the exception of Turin and Frosinone, that are manufacturing local systems (respectively of large enterprises and medium enterprises), the others are non-manufacturing local systems.
19. There are five North American and two French groups: 3M (US), Air Liquide (FR/US), Aventis (FR) Glaxo Wellcome (US), Heinz (US), Pharmacia (US/SV), Procter & Gamble (US). Foreign multinationals are here intended as the enterprises whose shares are entirely or mainly held by parent companies based abroad. The exceptions are the groups Artsana and Star, which are considered Italian here as their shares are held, through companies with headquarters abroad, by Italian families.
20. Turin, Bergamo, Desio, Milan, Voghera, Genoa, Castelmassa, Verona, Pordenone, Bologna, Fornovo di Taro, Parma, Ascoli Piceno, Siena, Rome, L'Aquila, Pescara and Brindisi. The presence of more than one foreign group is seen in four local systems: Milan, Rome, Verona and Frosinone. Only in the latter is the presence of Italian groups not found.
21. Furthermore foreign enterprises do not seem to prefer locations on the Italian border nearest to the home of the parent company. The two French groups, Air Liquide and Aventis, do not have preferences for the Piedmont region. Generally foreign multinationals groups seem to have a higher propensity to investment in the central, south and southern regions. This may be connected with a higher sensitivity of foreign MNEs to public incentives for localization in Italian depressed areas.
22. But other factors may be relevant as well. For example, MNEs are located in the local system of Pescara (in the Abruzzo region) which has a good index of specialization in

the pharmaceutical sector. Let us recall here the public incentives for the development of marginal areas; that is, the structural funds of which Abruzzo, together with other areas of European Union objective 2, has been a beneficiary in recent years.

23. The highest number of manufacturing plants belongs to the size comprising between 100 and 499 employees; just a few manufacturing plants have more than 499 employees.
24. Of course, different thoughts could be devoted to the effects (direct or indirect) on the size of the firms of the local systems where MNEs operate and on the structure of these systems.
25. The same results are found for MNEs of various sectors, in an investigation relative to 1993, still on the R&S database, concerning only local systems of small enterprise. See Tessieri (2000).
26. At this level of analysis we are not able to distinguish whether it is the presence of the multinational enterprises that determines some characteristics of the local system (the sectors of specialization and the prevailing size of the production units) or, on the contrary, the sectoral characteristics of the local system come, predominantly, from independent local factors that may be an important element for the location choice of the multinational enterprises.
27. Milan, Pavia, Verona, Padua, Parma, Ancona, Ascoli Piceno, Florence, Pisa, Siena, Rome, Latina, Catania.
28. All the specific cases of the HIBs of the 18 dedicated and non-dedicated multinational groups have been examined. Several cases of sector correlation (direct and indirect) have been found, especially for the HIBs of dedicated groups.
29. The groups are 3M Italia Finanziara, A. Menarini Industrie Farmaceutiche Riunite, Air Liquide, Aventis Pharma, Bracco, Compart, CSII Industrie, Heinz, Pharmacia & UP John, Snia.
30. Air Liquide Italia, Bormiolo Rocco & Figlio, Bracco, Compart, Finaf, Heinz Italia.
31. This is a 'non-dedicated' group, the Snia group, which through Bellco and Dideco operate in the local system of Mirandola.
32. Only two of the 19 local systems have a very high index of specialization in the biomedical sector, while eight have a very high index of specialization in the pharmaceutical sector.
33. We are referring to all of the Italian local systems.
34. Novi Ligure, Velletri, Savona, Isola della Scala and Reggio Calabria.
35. No local system in 2LS is characterized by the biomedical sector. Only two out of six companies appear to be connected to the first specialization of the local system where they have manufacturing plants. Moreover, only one out of six companies appears to be located in a local system with the pharmaceutical sector among the first three sectors of specialization. In addition, only one company is present in more than one local system, and only one multinational group is present with different companies in more than one local system.
36. Where the main MNEs that operate in the health industry or which could be indirectly or only potentially connected to it are not localized.
37. They include (a) all the local systems specializing in the health industry with at least one out of the two indexes of specialization in pharmaceutical or biomedical sectors, at the top or near the top of the list of all the manufacturing indexes of specialization; that is, at least in the third row; (b) those in the fourth, fifth or sixth row, if they also show one of the two indexes superior to 6.00. There is no definite rule for determining these thresholds.
38. Just one group ('non-dedicated') out of 18 of the first set operates in the local system of Siena, through only one manufacturing plant. It is the Bormioli & Figlio group. In the local system of Pisa we find the headquarters of two societies of the dedicated group, A. Menarini Industrie farmaceutiche Riunite.
39. One interesting case, hosting headquarters and plants of multinational groups, is Mirandola, where the index of specialization is very high. See Bellandi and Cencini (2001) for details and references on this case.
40. This last consideration emerges by analysing the *filière* connections of these MNEs within each local system: see Tessieri (2003).

41. That is, high-tech industries, traditional industries, high intensity of scale industries and specialized suppliers industries. Details on the correspondence with the three-digit classes of the 1991 ISTAT ATECO adopted in this chapter may be obtained from the authors.
42. Here 'innovative systems' have been defined as sets of nearby provinces oriented towards innovation in a specific technological class (data on patent activities, exports and foreign direct investments have been examined). See Ferrari et al. (1999).
43. There are also five HIBs plants in Milan and two in Rome. However, in neither Milan or Rome are there HIBs with both headquarters and manufacturing plants.
44. Among the innovative systems centred on Milan, those of plastic and fibres, chemistry, electronics and telecommunications, of precision instruments, measure and control apparatuses and, lastly, of electronic components may be recalled. These are sectors that could host activities connected to the health industry. In Milan, the multinational companies of the second type (that is, those which could have indirect connections with the health industry) mainly belong to these sectors. The 'innovative systems' centred on Rome are relevant also in the sectors of electronic components, of instruments of control and measure, and of refined chemistry products. All these innovative systems, except the biomedical ones, are anyhow of secondary importance compared to the ones in Milan. In Rome, the multinational companies of the second type mainly belong to these sectors.
45. Firstly, Mirandola, with its district-like structure, then Parma and perhaps Latina (which is, however, under the area of influence of Rome).
46. As the most frequently recurring within the list of sectors included in the set of sectors showing the three highest indexes of manufacturing specialization at the local level.
47. A typical example is the case of multinationals belonging to the fashion industries: Tessieri (2003).
48. Sanitary or ICT services are supplied by Bracco, Snia, Compart and Air Liquide, which operate in the metropolitan local systems of Milan, Genoa and Palermo, and in the non-manufacturing local system of Marsala.
49. In more detail, Milan is the Italian local system hosting the highest number of headquarters and manufacturing plants of MNEs within the 182 big firms of the R&S directory (see section 3), while Turin is the second, and Rome the third. Of course, the numbers and the relative positions change with the different industries. See Tessieri (2003).
50. Different choices, concerning location, degree and nature of embeddedness, emerge by comparing not only these MNEs, which comprise HIBs whose activities belong to different sectors characterized by different technological content too, but also MNEs that operate in different industries. For a comparison between the health industry and the fashion industry, see Tessieri (2003).
51. In order: Ceva (where a manufacturing plant of the Aventhis Pharma Group is located), Crescentino (two plants of the Snia Group), Frosinone (two plants of the Finaf Group) and Aprila (one plant of the Procter and Gamble Group), Cagliari (one plant of the Air Liquide Group), Ancona (two plants of the Finaf Group) and Siena (one plant of the Bormioli Group).
52. In this case MNEs could be interested in acquiring local units also belonging to their partners (for example, subcontractors) in order to absorb local resources and competencies.
53. However, signs of pure industrial cluster effects not necessarily connected with intensive local specialization in the biomedical or pharmaceutical sectors concern a larger number of local systems. These are the cases where the indexes of specialization in the biomedical or pharmaceutical sector are between 1.00 and 2.00, and at the same time intensive local manufacturing specialization in other sectors connected with the first ones is present locally. Moreover signs of a broader industrial cluster effect, not connected directly with the health industry, could have been considered. This is explained by the fact that many of these MNE groups are not dedicated to the health industry and the activities carried out by their HIBs belong to different Pavitt sectors with a different technological content too. For instance, some MNE groups operate for the food industry and their local units are in local systems where the manufacturing specialization in the food

industry is very high; other MNEs operate for the glass industry, and some of their local units are located in local systems where the specialization in the glass industry is very high; and so on; for more details see Tessieri (2003).

54. For instance, we may imagine the innovation systems (regional, national and international) in which MNEs could be involved.
55. Particularly: the scientific and/or academic milieux: see Cantwell and Iammarino (1998). On the localization of the research activity of the multinationals and the role of the incentives for investments, see Cantwell and Mudambi (2000). On firm-specific and location-specific characteristics driving organizational practice in pharmaceutical research laboratories, see Furman (2001). For investigations on the biomedical research in an Italian region (Tuscany) see Zanni (1996). On these and other related issues we have also referred to a wide bibliographic research in Tessieri (1997).
56. Supporting actions that make easier the investments in R&D for the pharmaceutical/biomedical industry; developing positive complementarity between different sectors, especially as regards the world of manufacturing industry and that of knowledge-intensive services.
57. Different types of clusters, by local systems, regions and nations, have to be spatially identified, analysed and compared in a systematic and dynamic way.

REFERENCES

Antonelli, C., 'Collective knowledge communication and innovation: the evidence of technological districts', *Regional Studies*, **6**, 2000, 535–47.

Bailey, D., Harte, G. and Sugden, R., 'Regulating transnationals: free markets and monitoring in Europe', in K. Cowling (ed.), *Industrial Policy in Europe*, London and New York: Routledge, 1999, pp. 311–25.

Becattini, G., *Il distretto industriale. Un nuovo modo di interpretare il cambiamento economico*, Turins Rosenberg & Sellier, 2000.

Becattini, G., Bellandi, M., Dei Ottati, G. and Sforzi, F., *From industrial districts to local development. An itinerary of research*, Cheltenham, UK and Northampton, MA, USA: Edward Elgar, 2003.

Bellandi, M., 'Local development and embedded large firms', *Entrepreneurship and Regional Development*, **3**, 2001, 189–210.

Bellandi, M. and Cencini, S., 'Biomedical firms in Italy's Mirandola high-technology cluster and their relationships with local and non-local health organisations', paper presented at the III Ferrara Health Industry Policy Forum, Ferrara (Italy), 2002.

Cantwell, J.A. and Iammarino, S., 'MNCs, technological innovation and regional systems in the EU: some evidence in the Italian case', *International Journal of the Economics of Business*, **3**, 1998, 383–408.

Cantwell, J.A. and Mudambi, R., 'The location of MNE R&D activity: the role of investment incentives', *Management International Review*, **40**, Special Issue 1, 2000, 127–48.

Cooke, P., *Knowledge Economies. Clusters, Learning and Cooperative Advantage*, London: Routledge, 2002.

De Propris, L., 'Diagnostics of local production systems', mimeo, Birmingham Business School, 2001.

Di Tommaso, M.R. and Schweitzer, S.O., 'L'industria della salute oltre il contenimento dei costi', *L'Industria*, **3**, 2000, 403–26.

Dicken, P., Forsgren M, and Malmberg, A., 'The local embeddedness of transnational corporations', in A. Amin and N. Thrift (eds), *Globalization, Institutions,*

and Regional Development in Europe, Oxford: Oxford University Press, 1994, pp. 23–45.
Enright, M.J., 'Regional clusters and multinational enterprises: independence, dependence or interdependence?', *International Studies of Management & Organization*, **2**, 2000, 114–38.
Ferrari, S., Guerrieri, P., Malerba, F., Mariotti, S. and Palma, D. (eds), *L'Italia nella competizione internazionale secondo rapporto,* Milano: Angeli, 1999.
Furman, J.L., 'Firm effects vs. location effects in the organization of pharmaceutical research', paper presented at the Academy of Management meetings, Toronto, August 2001.
Hood, N. and Young, S., 'Globalization, multinational enterprises and economic development', in N. Hood and S. Young (eds), *The Globalization of Multinational Enterprises, Activity and Economic Development*, London: Macmillan, 2000.
Istat, *censimento generale dell'industria e dei servizi*, Rome, 1991.
Istat, *censimento intermedio dell'industria e dei servizi*, Rome, 1996.
Istat, '*I sistemi locali del lavoro 1991*', a cura di F. Sforzi, Rome, 1997.
Keeble, D. and Wilkinson, F., *High Technology, Cluster, Networking and Collective Learning in Europe*, Aldershot: Ashgate, 2000.
Owen–Smith J., Riccaboni, M., Pammolli, F. and Powell, W.W., 'A comparison of U.S. and European university – industry relations in life sciences', *Management Science*, **0**, 2001, 1–19.
Pavitt, K., 'Sectoral patterns of technical change: towards a taxonomy and a theory', *Research Policy*, **6**, 1984, 343–73.
Porter, M., *On Competition*, Boston: Harvard Business Review Books, 1998.
R&S (Ricerca e Sviluppo), *Annuario*, Milan: Mediobanca, 2001.
Sforzi, F. and Lorenzini, F., 'Distretti industriali', in IPI, *L'esperienza italiana dei distretti industriali*, Roma, 2001, 20–33.
Solinas, G. and Baroni, D., 'I sistemi locali manifatturieri in Italia 1991–1996', in Becattini, G., Bellandi, M., Dei Ottati, G. and Sforzi, F. (eds), *Il caleidoscopio dello sviluppo locale: trasformazioni economiche nell'Italia contemporanea*, Turin: Rosenberg & Sellier, 2001, pp. 395–417.
Tessieri, N., *Multinazionali e distretti in Italia*, Tesi di laurea, Università degli studi di Firenze, 1997.
Tessieri, N., 'Multinazionali e distretti industriali in Italia', *Sviluppo locale*, **13**, 2000, 17–99.
Tessieri, N., 'Multinazionali e sistemi locali in Italia: alcuni risultati per l'industria della salute e della moda', doctoral thesis, Università degli Studi di Firenze, 2003.
Vaccà, S., *L'impresa transnazionale tra passato e futuro*, Milan: Angeli, 1995.
Vázquez-Barquero, A., 'Inward investment and endogenous development. The convergence of the strategies of large firms and territories?', *Entrepreneurship and Regional Development*, **11**, 1999, 79–93.
Zanfei, A., 'Transnational firms and the changing organisation of innovative activities', *Cambridge Journal of Economics*, **5**, 2000, 515–42.
Zanni, L., *Le imprese farmaceutiche operanti in Toscana: caratteri strutturali e dinamiche competitive*, Florence: CESVIT, 1996.

APPENDIX

BOX 11.1 SOURCES, LEGENDS AND TERMINOLOGY

Sources of Tables 11.1, 11.2, 11.3, 11.4: our elaboration on R&S (2001), Istat (1997), Solinas and Baroni (2001).

Typologies of Multinational Group: Dedicated MNE Group = multinational group with activities all connected to the health industry; Non-dedicated MNE group = multinational group with activities partially connected to the health industry; Only potentially connected MNE group = multinational group whose activities could be indirectly connected to the health industry, but of which we have no confirmation; Certainly not connected MNE group.

Typologies of Multinational Companies: HIBs = health industry branches; non-HIBs = non-health industry branches.

Typologies of Local Systems: LMLS = large firms manufacturing local systems; MMLS = medium-sized firms manufacturing local systems; SMLS = small manufacturing local systems; MLS = manufacturing local systems (LMLS + MMLS + SMLS); MELS = metropolitan local systems; NMLS = non manufacturing local systems (MELS + ONMLS); ONMLS = other non-manufacturing local systems.

HIBs and Sets of Local Systems: I set (1LS) = 50 local systems hosting HIBs; II set (2LS) = 5 local systems hosting non-HIBs; set of local systems specializing in the health industry (not included in 1LS or in 2LS) = 56 local systems with high pharmaceutical and/or biomedical specialization index; set of health industry local systems = 26 local systems that show a particularly high degree of specialization in the health industry.

12. High-technology clusters in France: two unusual models – an empiric study

Grégory Katz-Bénichou and Gérard Viens

INTRODUCTION

Over the last two decades, the development of businesses based on scientific and technical knowledge, and in particular those in the fields of information technology and biotechnology, has quite clearly been a driver of economic growth in North America as well as in Europe. The United States has established itself as the leader in this area by developing networks which link together universities, research centres and business centres. This is not the case in Europe, where relationships between the academic and business worlds have not been as widely developed, partly owing to laws that limit such relationships and partly because of a traditional reluctance to do so.

However, it is true that, since the 1980s, Europe has become much more interested in technological policy and in the transfer of technology from the academic to the business world. In a climate dominated by fears that the European Union is not competitive enough in industrial sectors dependent on scientific knowledge, an initial programme of R&D technologies was set up in 1987, after the Single European Act had been adopted (Pammolli et al., 2002). Such programmes sought to foster relations between universities and businesses by insisting on swift, tangible results which would lead to practical business applications. In spite of these changes, the United States has still maintained its dominant position ahead of Europe as far as innovation and transfer of technologies to the business world are concerned.

The usual explanations for this gap are well known. The first is that, over the years, the United States has done better than Europe when it comes to investing in small biotechnology enterprises with highly developed research activities. Secondly, there are disparities in state-sponsored R&D. A third reason is the positive role that incentives to encourage R&D have had. Here the Bayh–Dole Act of 12 December 1980 – a law that awarded the right to universities to take out patents on inventions which benefited from federal grants – had an influence in the United States. It allows companies to sell licences until they manage to raise funding from capital-risk investors by approaching consultants in intellectual property and universities specializing in technology transfer.

While being a true reflection of reality, this scenario is incomplete. Another main factor should be added, namely the wide variety of actors involved in R&D and the network structure that brings them together. Indeed, in order to develop and enhance technology transfer, very close links should be created between public research bodies, universities, technology start-ups and businesses. This is the type of set-up that gives birth to dynamic research. This type of organization explains why the American system achieves such high performances. It is resolutely decentralized, both geographically and economically. Indeed every public university receives funding from a wide variety of sources. There is, of course, public funding, but also that provided by private foundations, student fees, donations from alumni and revenues from industrial property. Significant research funds managed by the centres of the National Institute of Health (NIH) support interactions between pure research and clinical and pharmaceutical research. These different resources ensure that a wide variety of industrial applications ensue.

In Europe, funding takes place for the most part on a national level and each European country has adopted its own approach to the question of R&D. Without a single source of funding akin to that provided by the NIH, the search for grants assumes a competitive nature between laboratories, slowing down the business development of R&D. In some cases, national resources are spread out across a large number of small laboratories or are concentrated in one or two centres of excellence. So far the European programmes have only served to partially modify these characteristics.

Besides the problems of funding, the high-tech and biotechnology sectors may be considered to be examples of 'market failures'. Corporations involved in them have shown that they are not able to innovate enough, probably because the return on investment takes such a long time in such markets. It therefore follows that the state feels obliged to intervene by adopting an aggressive industrial policy. It is worth mentioning here the reasons often put forward to vindicate a state applying an industrial policy:

- protecting industries,
- preserving jobs and skills,
- fostering research in areas with strong growth potential,
- identifying the lack of competitiveness in international markets,
- ensuring that ethical standards continue to be reached,
- guaranteeing national solidarity.

Setting up and Developing Regional Poles of Excellence in Europe

The different approaches to change explored in this chapter are founded on a simple conviction. The obstacles in Europe – and particularly in France – are much too deeply rooted to imagine immediate and fundamental reform. This explains the interest shown in the dynamism of regional infrastructures organized in technology poles which are based on the American cluster models.

It might be worthwhile noticing that there is no uniformly accepted definition of a 'cluster' or a 'science park' in the literature. There are several synonyms used to describe similar developments, such as 'research park', 'technology park', 'business park' and 'innovation centre' (Monck et al., 1988). Westhead and Storey (1995) claims that science parks reflect the assumption that technological innovation stems from scientific research and that parks can provide the catalytic incubator environment for the transformation of 'pure' research into production. In this study, we consider that a technology pole is a rather spontaneous association of innovative businesses, of research and higher education institutions whose purpose is to promote cross-fertilization and job creation in advanced technologies (Faure et al., 1999). Technology poles have a European or an international vocation; they are specialized and must bring together universities, public and private laboratories and a network of companies.

The present study sets out to analyse two French high-tech clusters. One is Genopole at Evry in the Paris area, and the other is Futuroscope at Poitiers. Why have these two particular sites been chosen? The simple answer is that they serve as unusual examples, not only because of the way they were set up, but also because of the way they were financed. Genopole is a high-tech cluster specializing in biotechnologies. It stands out because the initial funding came from a charity whose purpose was to provide aid to patients with genetic illnesses. Since this initial stage, public funds have taken over and Genopole is expanding through a national network. Unlike Genopole, Futuroscope was launched by regional and local government authorities, in 1987. It did not position itself as a high-tech cluster in the beginning: quite the contrary; it has grown out of its success as a theme park tourist attraction and was inspired by the Epcot Center model in the United States. Futuroscope became well known and profitable, and this allowed the local authorities to fund and develop a high-tech cluster devoted to image technologies.

To measure the development of the two models, it has been decided to adopt given parameters: history and technological positioning, the origin of funding, job creation, property strategy, teaching and research, budget allocations, the existence of research laboratories and future projects.

However, it is clear that this approach has its limits. The most obvious one is that it will not give specific keys to compare performance *quantitatively* in terms of research and economic development. However, a *qualitative* analysis was chosen, which not only throws light on the subject from a French perspective, but also helps us understand how a European research space is evolving.

ENHANCING RESEARCH: THE REASONS FOR FRANCE'S DELAY

Since the mid-1990s, France has slipped behind, for two reasons. There is a chronic shortage of capital-risk funds and very heavy taxation. The biotechnology industry magnifies the general problem in France. In fact, in this sector the numerous small start-up companies, which usually operate in the health sector and devote a large proportion of their spending to R&D, embody the spirit of business innovation and enterprise. At this very moment, when many economists stress the importance of the complementary nature of public and private financing, the biotechnology field offers wide scope for experimenting, combining public fundamental research with company creation as well as public investment with capital risk (Kopp and Prud'homme, 2002).

Capital Risk

On this score, France is well behind its neighbours, Germany and the UK (Katz-Bénichou and Faure, 2001). The market capitalization of the sector is very low in France: four times lower than in Germany and three times lower than in the UK. There are fewer companies quoted on the Paris Stock Exchange. This was true not only in the year 2000, but also throughout the 1990s. The number of initial public offerings between 1993 and 2000 were twice as few in France compared to the UK and 1.7 times as few compared to Germany. A more significant fact is that French IPOs are worth half of those in the UK and a quarter of those in Germany. Moreover, capital risk, which makes up the vital resource to develop the sector, is lacking in France. See Table 12.1.

According to analysts from Ernst & Young, France ranks third in Europe for the development of its biotechnology sector (Table 12.2). The UK is first, distinguishing itself by a better market capitalization, a reflection of the maturity of its biotech companies. Next comes Germany, which heads the field in terms of number of companies, benefiting from a national and regional aid plan destined to help start-ups. According to

Table 12.1 Comparison between biotechnology industries in 2002

	UK	Germany	France	Europe	USA
General macroeconomic data					
Population (millions)	59	83	59	375	281
GNP (€ m.)	1 693	2 254	1 547	10 807	11 956
Biotechnology companies					
Staff	18 400	10 700	4 500	51 000	162 000
Number of companies	280	340	250	1 570	1 273
Total turnover (€ m.)	2 066	786	757	8 679	23 750
Financing of biotechnologies					
Number of IPOs, 2000	4	8	0	39	64
Total value of IPOs, 2000 (€ m.)	383	927	0	2 950	6 698
Average value of IPOs, 2000 (€ m.)	96	116	0	76	105
Total number of IPOs, 1993/2000	49	19	11	229	207
Average value of IPOs, 1993/2000 (€ m.)	39	67	18	111	262

Source: Kopp and Prud'homme (2002).

Table 12.2 Development of biotechnologies in Europe: comparative analysis, 2001

Country	No. of companies	Staff	Turnover (€ m.)
UK	280	18 400	2 066
Germany	340	10 700	786
France	240	4 500	757
Switzerland	110	5 600	1 313
Rest of Europe	590	21 900	3 757
Total	1 560	61 100	8 679

Source: Ernst & Young (2001).

this plan, for each euro provided by private funds, the state will provide another.

The year 2000 was very favourable for capital-risk investment in European biotechnologies. Germany increased this type of investment by 280 per cent, the UK by 72 per cent, whereas France increased it by only 34 per cent. Over the last decade, the net return on investment was between

15 and 25 per cent, the same as in France and the United States. Thus, even when France sees the advantages and the profitability of a strategy aimed at the development of biotechnologies, it does not marshal the necessary resources to match those of its main rivals. On top of that, the very high rate of taxation in France handicaps the sector even more.

Crushing Taxation

Compared to a German, an English or an American company, a French firm suffers from three handicaps. Tax on profit and salaries and on dividends distributed by companies is high, as is the tax on wealth creation. Table 12.3 shows this triple handicap clearly.

Tax systems are always complex and cannot be easily compared. The rules governing depreciation and deductions are not uniform and complicate

Table 12.3 The burden of taxation in France

		France	Germany	UK	USA
	VAT (%)	19.6	16.0	17.5	—[a]
Taxation related to the operation of the company	Tax on salary	60.2	40.8	21.9	15.3
	(%) Tax on profit	33.3	20.5	25[d]	30[b]
Taxation: earnings distributed by the company	Income tax (%) for the higher brackets	61	48	40	40[c]
Taxation: value added by the company	Value added on capital (%)	26	— (included in income tax)	20–40	20
	Inheritance tax (%)	5–60	7–50	0–40	—
	Wealth tax	Yes	No	No	No

Notes:

[a] Sales tax imposed by states, included in a range between 0 and 8%; cannot be compared to VAT rates.

[b] Federal rate of 15 to 35% plus a rate levied by each state ranging from 1 to 12%, depending on the state. This gives a total ranging from 16 to 47%; an average rate of 30% has been adopted for the purposes of this study.

[c] Federal rate of 35% plus a variable state rate between 1 and 12% depending on the state; 40% is a meaningful average.

[d] The rate ranges from 10 to 30%; we have chosen 25%.

Source: Kopp and Prud'homme (2002).

matters even more. Nevertheless the information shown enables us to appreciate overall how high taxation is in France compared to the other three countries. Thus researchers who become entrepreneurs, whose qualities and skills should command high incomes, enjoy net income after taxation that is lower in France, with the result that they are tempted to go and work abroad. Moreover the heavy burden of French taxation discourages foreign researchers from coming to France.

NEW MEASURES TO CLOSE THE GAP

The American Model as an Example

In the context of keen international competition in the field of biotechnologies, the Ministry of Research hopes to make France a more attractive place for researchers. It also aims to reverse the trend of researchers leaving for private industry. The reference model in economic terms remains the American one, characterized by a tradition of innovation and entrepreneurial risk.

In fact, academic research plays a very important role in the United States. The National Institute of Health (NIH) is one example. Its budget is 57 times bigger than that of INSERM, the French public Institute for Medical Research (Barre and Esterle, 2002). It has grown by at least 14 per cent a year since 1998 and by 16 per cent in 2002. Moreover ties between the world of academic research and biotechnology companies are very close indeed. The cultural and tax environment encourages the entrepreneurial spirit and values risk taking. This explains why capital risk funds are much more numerous in the United States. Lastly, the NASDAQ has enjoyed a growth rate of 10 to 15 per cent over the past two decades – excluding the period 2001–2. The rise has ensured that companies are financed and investors rewarded (Lenoir, 2002).

Considering the handicap that high taxation had imposed on the development of research by 1999, the French government decided to introduce reforms aimed at enhancing innovation and the creation of new businesses. It brought in a law to encourage scientists and researchers to set up their own businesses. A budget of €78 million was put aside for this purpose. The 1999 law on innovation and research (Law no. 99–587 12 July 1999, published in the *Journal Officiel*, 13 July 1999) addresses four issues.

1. *Increased mobility of researchers towards industry*. Public research does away with the barriers that prevent spin-offs and builds bridges between the state and private enterprise. In this way, faculty and university

researchers may stay within the state system and at the same time act as consultants in the private sector, set up their own companies, invest in the equity of a company and sit on the boards of private companies.
2. *Tax incentives and legal exceptions* for innovative companies.
3. *Development of patents in public organizations*, linked to universities and research centres.
4. *Creation of new business incubators* linked to universities and research centres (for example, the bio-incubator Paris Biotech links University Paris-V, INSERM, Ecole Centrale and ESSEC Business School).

In order to overcome its handicaps and enhance the value of its strong points, especially the quality of its science courses and the international reputation of its health system, France has taken the necessary steps to make sure that networks and regional poles of excellence will emerge in the realm of life sciences. Moreover the key to successful development of biotechnologies lies in the capacity to associate health and medical centres, scientific institutions and industrial concerns in an economic network and bring together start-up companies and the pharmaceutical sector. Following in the footsteps of the United States, Germany and the UK, which have all developed high-tech 'clusters', France is engaged in the process of developing specialist technology parks.

Before the advent of the 1999 law on innovation and research to ensure funding for innovation, and despite the scarcity of start-up funding, technology parks had succeeded in getting past the start-up phase and laid the foundations of long-term commercial development. Two examples serve to illustrate this phenomenon, the Genopole at Evry and Futuroscope at Poitiers. The sources of the capital for their development are rather unusual. Both parks owe their successful initial growth to their ability to benefit from the extensive media coverage they enjoyed. Their notoriety ensured that they were able to attract traditional funding from the public and private sectors, guaranteeing them long-term expansion.

Genopole, Launched Commercially with Capital from Charities

So as to make up for lost ground in the field of innovation and research and to foster the emergence of a real innovation culture, France devised a policy to set up a network of high-tech clusters specializing in the sciences of the genome. The originality of the project lay in its initial funding. In fact, at the outset, Genopole grew out of an initiative by a charity organization, The French Society Against Myopathies, which sought to collect funds for research into genetic illnesses. In 1987, it launched a television programme called *The Téléthon*, which broadcast for 30 hours non-stop and persuaded the general public to send money to help biomedical research. Since then media coverage has continued to be very wide and, in 2004, the initiative brought in the sum of €98 million (www.afm-france.org).

Using capital from the Téléthon, The French Society against Myopathies (AFM) set up the Genethon laboratory in 1990. It was to publish the first physical genetic map of the human genome (Daniel Cohen and Jean Weissenbach) three years later. The first initiatives undertaken by the AFM and Genethon to create a 'Genome Valley' at Evry date back to 1994. The following year, the Genset company was set up in Evry and launched the programme known as 'Great Sequencing'. At this early stage of its development, the Genopole project relied solely on financing from charity and private funds. Then, in 1996, INSERM, the French national institute for medical research, highlighted the opportunity for France to create a national centre for sequencing and genomics. It was inaugurated in 1997.

In 1998, the Minister of Research of the time, Claude Allègre, appointed Pierre Tambourin, the former Director of the Life Sciences Department of the National Research Institute (the CNRS), to coordinate and manage Genopole. In 1999, the technology park became a not-for-profit organization and the minister decided to set up a network of eight Genopoles in French cities, based on the Evry model.

The Property Strategy of Genopole

The budget invested in building and converting premises amounted to €3.5 million. Funding came from the state and local authorities. In 2001, Genopole rented 15 000 sq.m. in Evry to house laboratories and companies. Given the lack at Evry of premises suited to the purposes of research, Genopole decided to keep the prime premises for public and private research activities and rent them out to start-ups or laboratories. They chose this approach in order (a) to dispose of premises suited to the specific requirements of laboratories at market price, (b) to make the organization of space a priority so as to develop a campus-type organization in the long run, and (c) to set up services for tenants as required, encouraging the sharing of scientific equipment, secretarial and consulting services and sanitary services for the laboratories. This sharing of its activities resulted in savings in operating costs.

At the same time the local authority set up an organization, '*Territoire Genopole*', in order to measure the impact of the project on the environment of the city and the surrounding area. In 1999, the Minister of Research decided to create a national network of high-tech clusters specializing in genomics in eight French cities (Evry, Lille, Strasbourg, Bordeaux, Lyon, Montpellier, Toulouse, Marseille) for which the Genopole in Evry was to be the pilot model. Today the Evry Genopole includes 20 laboratories and 39 companies employing no fewer than 1500 people. Each centre is supposed to become a specialist in a specific area of genomics. This would ensure that the project would cover the widest number of research areas. Since 1999, the French state has put €69 million into the creation and development of this network.

The legal status chosen (an Economic Interest Grouping) will ensure that the fledgling network will become a long-term reality in economic and human terms.

Sources of Financing and Job Creation

In 2002, 39 companies operated in Evry. The declared goal of Genopole is to create a biotechnology cluster with 60 or more companies within two years. To reach this objective an 'incubator' has been set up in the National Centre of Genomics (with the support of the regional authority). Its purpose is to bring closer together public research and the business world. Results have been outstanding, as six companies have already raised €25.8 million. By July 2001, the incubator at Genopole had dealt with 21 projects.

The *Genavenir* programme aims to house start-ups in the vicinity of universities conducting research. At the same time a seed investment fund, '*Genopole 1st Day*' was set up to finance the emergence of new biotechnology enterprises. Fifteen private and public investors put up the fund's capital of €1.2 million. The main investors in the fund were AFM and Dassault Développement (20 per cent each). For each project the initial investment by the new company's founder (€15 200) is matched by an average investment from the fund of €45 700. This seed capital qualifies the company for the award of public grants, especially through ANVAR (The French national agency for the transfer of research to industry).

The various programmes and incentive schemes set up to attract industries and research laboratories to the site have led to the creation of new jobs. In 2001, the 1562 jobs on site comprised 240 in administration, 602 in companies and 720 in research laboratories. The aim for 2003 was a total of 5000 jobs on site, with 1000 researchers, 1500 students and 60 new companies.

This rather high growth rate in job creation on the Evry site serves as an example to the rest of the network. It is of the utmost importance for the technology parks to reach a critical mass as quickly as possible.

Promoting a Pole of Excellence in Teaching and Research

Measures have been introduced as incentives to attract researchers to Genopole. To persuade researchers to come and work on the site, Genopole makes sure that it offers them attractive working conditions. One way to achieve this goal has been to build a common technical platform, with €1 million invested in heavy equipment in 2001. Sharing the research equipment provides the opportunity to achieve economies of scale.

Genopole aims to reduce the brain-drain by enabling young postdoctoral students who are abroad to return to France. In October 2000, there were 587 people working on research; in spring 2001, the number had grown to 720. Local authority grants are offered to attract researchers. In 2001, 20 research laboratories were set up on the site. Today, the opening of three laboratories a year would be sufficient to ensure a reasonable rate of expansion and reinforce this Evry high-tech cluster. This would allow it to reach a critical mass and to make its presence felt on the international scene.

Laboratories at Genopole specialize in genomics, proteomics, biocomputing, robotics and nanotechnologies. Most of these laboratories have grown out of state research bodies such as INSERM, INRA, the CNRS and CEA. Genopole seeks to reach a very high level of research. Its credibility and its ability to attract interest will depend mainly on the quality of the research teams in place. That is why Genopole is aiming to become a recognized pioneer in biotechnology research at national and European levels.

To stimulate the growth of Genopole, communication has been fully integrated in the strategic development of the cluster. The purpose of 'Genopole Communication' is to communicate inside Genopole (seeking to federate the campus members around a common project) and with the outside world, regionally, nationally and internationally. The strategy sets out (a) to involve the actors on the site in the public debate about the life sciences, (b) to stage national and international conferences, and (c) more broadly, to promote and develop French research abroad.

A range of communication tools have been developed, which includes weekly press releases, conferences, seminars, breakfast meetings, publication of books and communication with the public at large. Genopole helps students in their search for internships through a specific structure which allows them to approach companies. It has also become part of the European network project, 'Bio Valleys'. This should make it an attractive place to return to for French postdoctoral students who work abroad.

As Genopole is concerned with spreading the notoriety and influence of the park, it has joined the European Union's technology research and development programme, called PCRDT, as well as a network of European bio-incubators. On the local level, the links with the University of Evry Val d'Essonne have become closer. The university had almost one thousand students enrolled in 2001 and has rapidly expanded its courses in biology, genetic engineering and information technology. In management studies, the focus has been on courses with a practical, professional emphasis. Postgraduate courses in business creation, entrepreneurial engineering and local economic development have been set up.

Budget and its Allocation

Tables 12.4 and 12.5 show that almost 87 per cent of funding for the technology parks comes from the local authorities (The Regional Council of the Ile de France and the Essonne Local Authority Council). As for expenditure, nearly 80 per cent goes towards the material cost of setting up the laboratories and operating costs. If the successful launch of Genopole was due to the AFM and the local authorities, today it is national and regional funding that ensures most of the financing of the public laboratories present on the site.

Table 12.4 Financial aspects, 2000

Resources (€ m.)		Expenditures (€ m.)	
National public funding	0.91	Premises	4.18
Regional funding	6.79	Equipment	1.58
Local authority funding	2.07	Grants	0.34
INSERM	0.03	Research projects	0.69
City council	0.15	Scientific collaboration	0.40
AFM (charity)	0.26	Communication, studies, conferences	0.55
		Operational budget	2.22
		Sundry expenses	0.25
Total	10.21	Total	10.21

Source: Annual report on Genopole.

Table 12.5 Total investment in the Genopole cluster, 2000

Source of funding	In millions of €	%
State	62.16	72.0
Local authorities	11.90	13.8
AFM	12.44	14.2
Total	86.50	100

Future projects

Genoplante is a programme in plant genomics, which is the result of a partnership between the public sector (INRA, CNRS, CIRAD, IRD) and the private sector (Biogemma, Bioplante). In 1999 the project had a budget of €213 million over five years.

GenHomme is a post-genomics human programme set up to coordinate the work of public laboratories, charity organizations and businesses. Its goal is to facilitate the development of knowledge in the field. The French state has awarded the programme a budget of €152 million over five years.

Strengths, weaknesses and future prospects

Among the *strengths*, the very high quality of the laboratories and researchers on the site represents the main factor of success. Although it has been supplanted to a great degree by state and private funding, the dynamism of the charity organization AFM did provide the initial impetus that launched the project. At the same time, the Téléthon gave it valuable national media coverage. Another strength is the synergy between developing the technology park and the new university at Evry, with the creation of the 'campus effect'. In the same way, having sponsors and partners helps to promote prestige and gives the park a strong brand image despite its being a recent creation. Its location near Paris makes it the ideal candidate as coordinator of the national network.

The main *weakness* is related to the fact that they are ten years behind biotechnology clusters in the United States, the UK and Germany. The development of venture capital funds in France and the recent law enhancing public research are beginning to show results. Whether this enables France to catch up with Germany and the UK remains to be seen.

There are two threats which risk slowing down the development of the Genopole network. One is that the expansion of high-tech clusters in the United States, the UK and Germany may attract French researchers who are looking for recognition of their work. The other is that France has some of the most restrictive laws governing bioethics. Public and political debates reveal distrust of GMOs, patents, stem cells, cloning and experiments with human embryos. These ethical concerns challenge breakthroughs in some techniques that other countries agree to explore, with the result that French researchers are attracted abroad.

As regards opportunities, Genopole is the result of the development of close links between public research bodies and private business in a sector that enjoys a high growth rate. The change in legal status, making it an Economic Interest Grouping with financing from both the public and private sectors, is a real opportunity for growth. The communication strategy adds to its notoriety and its attraction for teaching, research and industry. The incubator is part of the European Union's technology research and development programme and it also belongs to a network of European bio-incubators. This gives it an international reputation and leads to harmonization of budgets in research across Europe.

Futuroscope, Developing a Technology Park out of a Theme Park Attraction

Among the French technology parks, Futuroscope is a particular case. In 1984, the French government started to implement the law on decentralization. At the time, the Vienne district, which is part of the Poitou-Charente region, was poorly developed, with 370 000 inhabitants tempted by rural exodus. The president of the local authority sought to reverse the trend by propelling the region into tomorrow's world. He wanted to set up an original concept based on the Epcot model in Florida by bringing together under one roof the four main activities in human life: leisure, work, research and training. The Futuroscope theme park project was launched a little while later, in 1986.

The Epcot Center was very much a technological and scientific showcase set in the middle of a theme park. Unlike the American venture, Futuroscope did not have such a showcase. The creation of Futuroscope involved, however, an audacious undertaking, that of developing a theme park based on the technology of the image. It was an easy thing to build the infrastructure and the park attracted visitors, which helped spread its notoriety; above all, its development was a profitable business. The theme park was the financial lever that ensured the development of business on the site. The key idea was to develop, in the first stage, a showcase for the region, and then to build the university and research complex, which would allow the technology park to develop in spite of being in a rural area where resources are derived mainly from agriculture.

Some companies began to set up at the same time as the park was being developed. They felt isolated – a feeling that began to be called the 'potato field' syndrome. In 1992, the park started to become highly successful and the hotel industry began to grow. Futuroscope became well-known very swiftly and attracted an increasing number of visitors. At the same time the park had a positive effect on the region, as the population increased to 400 000 in 2000, while in the regions around it there was a decline.

Property Strategy

Until 1995, the park was a breeding ground for SMEs. Then demand for business premises changed completely. Bigger companies needed large premises in the surrounding area. The Cegetel group decided to settle in the high-tech cluster. In 1999, the property development programme for business premises covered 25 000m^2. Three years later, the figure was almost 40 000m^2.

The district adopted an astute real estate strategy which involved buying farmland close to the site so as to have land at its disposal for future development. As long as the land was not needed for the Futuroscope project, farmers were allowed to use it. Thus, by anticipating the future needs of the park while securing income from leasing the land back to the former owners, the district was able to make substantial savings. Access to the cluster was felt to be of major importance, so a railway station for the French high-speed train – the TGV – and a motorway were built serving Futuroscope. The capital to finance these two projects, which cost €15 million, was raised without the help of the state.

Some 15 years later it is clear that the cluster is a success. It exists today through the brand name the park has built up. This reputation, based on the development of tourism, has made it possible to develop the teaching–research activity and the high-tech cluster. It has acted as a powerful lever, attracting companies to the site.

Job Creation

The county has always been a farming region with no industrial or scientific tradition. In order to create one, the council was convinced it had to develop a project based on future technologies. This meant fostering a culture of innovation and promoting an entrepreneurial and business culture in order to increase the number of jobs in the region. To meet these challenges, several initiatives were taken to boost development. A business incubator funded by the council and the European Social Fund was set up to encourage the growth of local companies. By 2002, the number of jobs on the site had grown to 8000.

To attract companies to the Vienne, the district developed a network to find jobs for the partners of company employees who transfered to the area. Financial incentives were given to companies in order to encourage the hiring of younger staff. By 2001, the 8000 jobs on the Futuroscope site comprised 2100 students, 1700 teaching and research posts, 2500 company jobs and 1700 staff in the theme park and hotels.

Teaching and Research Activities

Close to the site, a university department specializing in teaching and research in mechanical and aeronautical sciences is being developed. There are almost 700 researchers from the French National Research Institute, the CNRS, at the university of Poitiers. This represents 10 per cent of researchers in engineering sciences in France. There is also ENSMA, an engineering school with several research laboratories specializing in aeronautics and space research. Although this concentration may be considered an advantage, this highly specialized field, mechanical engineering and the aeronautic industry, offers limited opportunities for innovation.

Only one laboratory specializes in new technologies in communication and information (image compression, signals and so on), but a large number of companies involved in the field have come to the cluster to develop their businesses.

Financial Aspects and the Management of the Park

Since the objectives of Futuroscope (the end of the rural exodus and development of scientific and industrial activities in the area) were achieved in 2000, the county sold the operation of the park to a privately-held French theme park company. The transaction was agreed on the basis of the park's annual turnover of €90 million. The amount agreed was half of the park's annual turnover. The council retained the ownership of the park.

Interestingly enough, owing to poor handling of the management of the park, the company in charge is now considering handing back the operation to the regional council. Total accumulated investment by the local authority since Futuroscope was set up in 1986 can be broken down as follows: the park, €309m.; the high-technology cluster, €54m.; training and research, €76m. Total cost of the project since the beginning is €435m.

It is not easy to assess the return on investment of the park. The benefits for the area exceed the financial results. The park has completely changed the region while reversing the rural exodus. The district is the only one in the region to have made the change from a rural economy to one of high-tech industries and services. Despite the recent management difficulties, the change has been an overall success. One of the factors that enabled the council to make this change was the speed of the decision processes due to the autonomy that it enjoyed. It decided to remain independent, rejecting offers and proposals from potential partners and proposals to form strategic alliances.

Project and Failure

In 2003, a theme park called Naturascope was about to set up, focusing on sustainable development and natural disasters. It was based on the Futuroscope model and was supposed to create a 'win–win' type of situation. While Futuroscope has developed in one field, that of mechanical engineering, Naturascope – concerned as it is with man and nature – offered a cluster of opportunities for development in the biotechnology field. As it would be located 30 kilometres from the Futuroscope site, it would increase the number of visitors and make the area more attractive for future investors, with the result of achieving higher inward investment. However, designed to be situated at the frontier of a protected forest with endangered species, the project was finally cancelled in 2004 because of strong protests from farmers and local associations (ATTAC).

Strengths, Weaknesses and Future Prospects

The main *strength* of the Futuroscope technology cluster is its excellent brand image, due to the theme park. The property strategy adopted to expand the park in a short period of time was proved judicious. Having a high-speed train station at the site was another shrewd move as the 330-kilometre journey from Paris takes only one hour and a half.

The main *weakness* is that the park has not yet reached the size that successful technology clusters usually do. Because of the highly specialized field it has chosen, the prospects for attracting new research laboratories are limited. It still suffers to a certain extent from its provincial image, which means that it remains difficult to attract any large French company. It is also true that the incubator has not yet been fully developed. On the whole, the economic and technological development has not been as successful as the theme park.

There are several threats regarding the future of the cluster. Firstly, the council having to take back control of the park two years after selling it shows the fragility of the operation and that the condition for success is innovative and reliable management. Secondly, the cluster is suffering from a lack of European exposure, which handicaps its development. Finally, the project has relied mainly on the vision and the commitment of one man and his team. It is vital to find new leadership in order to lay the foundations of long-term success.

CONCLUSION

The purpose of this chapter was to illustrate the state's industrial policies in the fields of high-tech industry and biotechnology. In order to cope with market failures in this sector, one answer in France, and indeed in other European countries, has been to help to develop high-tech clusters. The output of these clusters and the success of these industrial policies have to be measured within the framework of a multi-product production function, as suggested in the Health Industry Model of Di Tommaso and Schweitzer.

More specifically, the purpose of this study was to underline the creation of two high-tech clusters through models that are unusual both in their financing and in their development. They are remarkable for their specific differences rather than for their similarities.

While Genopole was a national project initiated by a charity organization, and whose development was ensured by government aid, Futuroscope was a regional project which was inspired only by local drive and ambition. The primary purpose of Futuroscope was to stimulate growth in the local economy: the cluster associated with the theme park was only a means to that end – an unusual yet successful one. Research did not appear, at the beginning, to be a priority. In the case of Genopole, it was the other way around. The main purpose was to enable French research in biotechnologies to catch up with other countries. The economic and business development of the high-tech cluster was the means of promoting research. On this point, the means and the ends of the two models were diametrically opposed.

Technological development and enhancing research rely on men, ideas and capital. However, in both cases, it was the wide media coverage that was a key factor in finding the capital. This also shows that funding for such big research projects does not necessarily rely on governments. Although these two approaches are not the only ones, they do show that there are alternative ways to the traditional American one of matching private and state funding. Creating events or activities which appeal to the public at large and building up a brand image through wide media coverage are also a strategy for raising the initial funds to launch a technology cluster.

Despite the differences and the specific nature of the challenges they have had to face, Genopole and Futuroscope seem to be lasting successes. Without doubt, the main reason for this is their one common point: both projects have proved capable of bringing together teaching, research and the corporate world. The link between business and university has stimulated job creation; research and business complement one another, stimulating and promoting technological innovation; and, lastly, the synergy between training and research has fostered both the expansion and the sharing of knowledge.

Looking to the future, to facilitate European research development it is important to understand the best practices in member states and emphasize the most suitable according to specific criteria. In its own way, the present study has shown two unusual clusters which will add to the wide variety of models currently implemented in different European countries.

REFERENCES

Bakouros, Y., Mardas, D. and Varsakelis, N., 'Science park, a high tech fantasy: an analysis of the science parks of Greece', *Technovation*, **22**(2), February 2002, 123–8.

Barre, R. and Esterle, L. (eds), 'Indicateurs 2002, Rapport de l'Observatoire des Sciences et des Techniques', *Economica*, 2002.

Blumberg-Morki, M., 'Biotechnologies: une strategie pour l' Europe, *Biofutur*, **222**, May 2002.

Brouthers, K. and Werner, S., 'R&D mode choices in Central and Eastern Europe', *Journal of Business Research*, **52**(1), April 2001, 83–91.

Busquin, P., 'Le fossé Europe Etat-Unis se creuse', *La Recherche*, June 2002.

Busquin, P., Jacob, F., Montagnier, L. et al., *Quel avenir pour la recherche?*, Paris: Flammarion, 2003.

Cueni, Thomas, 'The research-based pharmaceutical industries in Europe', *Pharmaceuticals Policy and Law*, **1**, 1999, 37–47.

Ernst & Young, 'Life Sciences in France', 2001.

Faure, P., Vidal, H., Pollard, H. et al., 'Genopole d'Evry et l'expérience de 10 technopolis français', *Genopole Annual Report*, March 1999.

Kanavos, P., 'A prospective view on European pharmaceutical research and development', *Pharmacoeconomics*, **13** (2), February 1998, 181–90.

Katz-Bénichou, G. and Faure, P., 'Benchmarking of biotech economic developments: a comparative study of France, Germany and United Kingdom', ESSEC Business School, Cergy, 2001.

Kihlgren, A., 'Promotion of innovation activity in Russia through the creation of science parks: the case of St. Petersburg', *Technovation*, **23**(1), January 2003, 65–76.

Kopp, P. and Prud'homme R., 'Dépénaliser l'innovation', Objectif, 2010 and France Biotech, discussion paper, 26 March 2002.

Lenoir, N., 'Relever le défi des Biotechnologies', a report to the French Ministry of Economy and Finance and the Ministry of Research, 11 March 2002.

Löfsten, H. and Lindelöf, P., 'Science parks in Sweden – industrial renewal and development?', *R&D Management*, **31**(3), 2001, 309–22.

Löfsten, H. and Lindelöf, P., 'Science parks and the growth of new technology-based firms, academic–industry links, innovation and markets', *Research Policy*, **31**(6), August 2002, 859–76.

Monck, C.S.P., *Science Parks and the Growth of High Technology Firms*, London: Croom Helm, 1988.

Mothe, C. and Quelin, B., 'Creating competencies through collaboration: the case of Eureka R&D consortia', *European Management Journal*, **18**(6), December 2000, 590–604.

Pammolli, F., Riccaboni, M., Owen-Smith, J. and Powell, W., 'Europe/Etats-Unis: la force de réseaux', *Biofutur*, **222**, May 2002, 26.

Shin Dong-Ho, 'An alternative approach to developing science parks: a case study from Korea', *Regional Science*, **80**, August 2000, 103–11.

Storey, D.J. and Tether, B.S., 'Public policy measures to support new technology-based firms in the European Union', *Research Policy*, **26,** 1998, 1037–57.

Westhead, P. and Storey, D.J., 'Links beween higher education institutions and high technology firms', *International Journal of Management Science*, **23**(4), 1995, 345–60.

Index

Titles of publications appear in *italics*.

Abernethy, M. 132
academic research, *see* universities, research
acceptable level of health 21–2
acquisitions, biotechnology sector 139–40
ageing population, and health service demand 34–6
agglomeration, *see* clusters
Albino, V. 185
Allio, M.K. 163
ambulatory health provision 28–9
Anderson, R. 22
Anglo American, strategic decision-making 51
anthrax, and strategic decision-making 51, 52
antidepressants 83
  hedonic pricing model 87–8
  therapeutic effectiveness measurement 85
antihistamines 83
  hedonic pricing model 91
antimicrobials 83–4
  hedonic pricing model 89–90
  therapeutic effectiveness measurement 85
Arrow, K. 128
Arthur, J. 132
Audretsch, D.B. 190
Australia, university research policy 116–19

Barrett, P. 106, 107
Bayh–Dole Act 143, 258–9
benchmarking hospital costs, UK 150–65
Bianchi, P. 63, 176
biomedical machines, innovative systems, Italy 245
biomedical research 135
biotech industry 136–40
  capital risk investment, Europe 261–3
  clusters 189–96, 208–17
    San Diego 227–8
  intangible assets 140–41, 177–81, 190–95
  startups 192–4, 208–9
  university links 209–10, 213–16
  see also pharmaceutical industry
Biovalley, biotech firm clusters 194–5
Bothwell, M. 177
Bottazzi, G. 139
branded drugs, pricing 81–2, 85–6
  antidepressants 87
  calcium channel blockers 88–9
Britain, *see* United Kingdom
Buigues, P. 125
Burton-Jones, A. 183

Caddy, I. 106
calcium channel blockers (CCBs) 83
  hedonic pricing model 88–9
California, regional spillovers 226–9
Camp, R.C. 153
Canada, pharmaceutical prices 37
capital costs, impact on benchmarking 161
capital risk fund shortage, France 261–3
casemix, impact on hospital benchmarking 161
centre-based approach to collaborative research 113–14
clusters 206–17, 224–6, 234–5
  biotechnology industry 189–96, 208–17, 227–8
  high-tech
    France 260–76
    Italy 235–49
  and intangible assets 182–96

Coase, R. 50
codified knowledge 184
collaborative agreements, biotechnology 138–9
collaborative research, *see* research alliances
collective learning, biotech companies 191
commercialization, university research 116–19; *see also* research alliances
communication, Genopole 268
competition
health plans, USA 24–5
international, health systems 37
competition–cooperation 72–3
complementarity 130–31, 134, 141–3; *see also* intangible assets
Consorzio Ferrara Ricerche (CFR) 120
corporatization of research, *see* research alliances
cost containment policies 22–5, 60–61, 62–5
costs
drugs, *see* pharmaceuticals, pricing
medical R&D 33
county manager view of firm location 210
Cowling, K. 50, 53

Danby, M.R. 209
decision-making, strategic, and economic impact 50–53
demand, health services 34–6
demand-related horizontal interactions, clusters 225
Desrochers, P. 184
development of economies 48–9
Di Tommaso, M.R. 46, 47, 171
direct business funding, collaborative research 114
direct intellectual capital methods (DIC) 173
drugs, *see* pharmaceuticals
Duffy, J. 105

East Asia, university research 109
economic development
and health status 10
of localities 48–9
economic efficiency, and industrial promotion 4
economic impact
and strategic decision-making 50–53
University of California research alliances 227–9
Economic Report of the President, US 6
economics and intangible assets 127–31
economies of agglomeration 208; *see also* clusters
Edvinsson, L. 108, 174
efficiency, health supply 38–9
employment
Futuroscope 272
Genopole 267
health sector 19–21
impact, UC San Diego research alliances 228
Enthoven, A. 65
entrepreneurial view of firm location 210
Epple, D. 82
Europe
biotech clusters 193–5
health policies 59–74
pharmaceutical industry 135–40
R&D funding 259
technology policy 258, 259
university research 109
Eustace, C. 174
expenditure, *see* health expenditure
export potential of health industry 36–7

Feldman, M.P. 185
Ferrara
economic development 48–9
University research activities 120–22
Ferrera, M. 71–2
finance
Futuroscope 273
Genopole 267, 269
health care 9, 29–30, 55
see also health expenditure
intangible assets 187
NHS, Britain 55
organisations 29–30
university research 110, 114–15, 116
firm's value, intangible determinants 133–4

foundation hospitals, Britain 55
France
  capital risk fund shortage 261–3
  health research 142
  high-tech clusters 260, 265–76
  research problems 261–4
  taxation 263–4
French Society Against Myopathies, The 265–6
funding, *see* finance
Future Management of Crown Copyright, The 107–8
Futuroscope 260, 271–5

G-7 countries
  ageing population 35–6
  employment, health sector 19–21
  expenditure, health sector 18–19, 22–9, 30–33
Gallagher, M. 117
Garcia-Ayuso, M. 132
generation of drugs, effect on pricing 86, 95
  antidepressants 87
  glaucoma treatments 92
generic drugs, pricing 81–2, 85–6
  antidepressants 87
  calcium channel blockers 88–9
GenHomme 270
Genoplante 270
Genopole 260, 265–71
geographical proximity, *see* clusters
Geroski, P. 127
glaucoma treatments, *see* opthalmic solutions for glaucoma
GlaxoSmithKline, strategic decision-making 51
globalisation of health issues 56
Goldsmith, J. 32
governance, health industry 55–6
government
  funding, *see* public funding
  policies, new technologies 12
  strategic decision-making 51–2
Great Britain, *see* United Kingdom
Guthrie, J. 118

Haig, R.M. 182
Hall, S.J. 182
Health Behavior Model 22
health care funding 9, 29–30, 55
health care providers 7, 9
  Health Industry Model 25, 27–9
health care resource groups (HRGs) 151
health expenditure
  G-7 countries 18–19, 22–9, 30–33
  hospitals, benchmarking 150–65
  management 6–7, 63–73
  see also health care funding
health industry 6–11, 46
  costs, *see* health expenditure
  decision-making 53–5
  export potential 36–7
  impact on economics 10–11, 47–53
  intangible assets 177–81
  structure 7–9
health industry branches, MNEs, Italy 238–40
  and local systems 240–46
Health Industry Model 7, 25–33, 46, 47, 207
  policy implications 33–9
health insurance
  role in health process 65–6
  USA 29–30
health maintenance organisations and drug prices 79–80, 94
Health Model 7, 21–5
  policy implication 33–9
health product manufacturers 9, 30–33; *see also* pharmaceutical industry
health service demand 34–6
  Europe 62–3
health systems
  international competition 37
  strategies, Europe 60–74
hedonic theory, drug pricing 80–97
Henderson, R. 135, 138
HIBs, *see* health industry branches
high-tech clusters 225
  France 260–76
  Italy 235–49
  see also biotech industry, clusters
high-technology industry promotion 3–6
higher education institutes, *see* universities
HIM, *see* Health Industry Model
HM, *see* Health Model

HMOs (health maintenance organisations) and drug prices 79–80, 94
Holloway, J. 153
horizontal interactions, clusters 225
hospitals, 25, 27–8
  costs, benchmarking 150–65
  as election issue 52
  foundation 55
HRGs (health care resource groups) 151
human capital as intangible asset 130–31
Hutton, W. 54

Independent Kidderminster Hospital and Health Concern Party 52
industrial district development 233–4; *see also* clusters
industrial health cluster effect 246
industrial policy, high-tech industry 3–6
industrial promotion 3–6
industry, research collaboration with universities, *see* research alliances
information 183–4
innovation
  impact of benchmarking 165
  as intangible asset 127–8
  law, France 264–5
  pharmaceutical, impact of pricing 77–8
  see also research and development
innovative systems and health industry, Italy 245–6
insurance, *see* health insurance
intangible assets 125–43, 171–96
  classification 174–7
  and clustering 182–96
  health industry 134–41, 177–81
intellectual capital 103–8, 131–2
intellectual property (IP) 103–8
international trade, health industry 37
intervention policies, Europe 60–73
investment, research and development 32–3, 172
Italy
  multinational enterprises and high-tech clusters 235–50
  university research policy 119–22

job creation
  Futuroscope 272
  Genopole 267

Kidderminster Hospital, as election issue 52
knowledge 183–4
  as intangible asset 128–9
  management 105–8
knowledge capital 132
  indexes, health companies 177–9

Labory, S. 176
'ladder of success' benchmark 155–7
LeGrand, J. 61
Lev, B. 132, 174, 176, 177, 184
Levin, R. 129
'*List of shame* to feature NHS Trust costs' (*Financial Times*) 152
local systems, Italy 241–4
  and MNEs 233–50
localisation, *see* clusters
Lowendahl, B. 174, 188

Maciocco, G. 66
macroeconomic intervention, Europe 60–61
managed competition policies 65–8
managed cooperation model 68–70
manufacturers, health products 9, 30–33; *see also* pharmaceutical industry
manufacturing local systems (MLS) 237
market capitalization methods (MCM) 173
market factors and drug pricing 79, 85–6
market failures
  ideas market 128
  rationale for industrial policy 4
market incentives to improve efficiency 38–9
market, international, health systems 37
market-to-book ratio 178–9
marketing, pharmaceuticals 33, 78–9, 180–81
Marshall, A. 224
Martin, M. 104

measurement, intangible assets 173–4
medical equipment 31–3
medical services, international trade 37
medical training and strategic decision-making 51–2
mergers, biotechnology sector 139–40
meta-economic objectives, rationale for industrial promotion 4, 6
microeconomic intervention 61, 65–8
Milgrom, P. 130
MMR vaccine, parental decision-making 53
MNEs, *see* multinational enterprises
Monsanto Co., and Washington University 221–2
Mossialos, E. 61
multinational enterprises (MNEs)
  and high-tech clusters, Italy 235–50
  and industrial district development 233–4

National Health Service, Britain
  managed competition model 66
  strategic decision-making 54–5
national reference cost index (NRCI) 151, 155–65
national reference costing exercise (NRCE) 151–65
National Schedule of Reference Costs (NSRC) 155–65
Naturascope project 274–5
Neumann, R. 118
new biotechnology companies (NBCs) 190–91
NHS, *see* National Health Service, Britain
non-manufacturing local systems (NMLS) 237
NRCE (national reference costing exercise) 151–65
NRCI (national reference cost index) 151, 155–65
NSRC (National Schedule of Reference Costs) 155–65

opthalmic solutions for glaucoma 84
  hedonic pricing model 91–2
organization as intangible asset 129–30
outpatient surgery 29

Paci, M. 71–2
Pammolli, F. 191
Paris, biotech cluster 194
paternalistic reasons for industrial promotion 4, 6
patient information 33
pharmaceutical industry
  intangible assets 134–41, 177–81
  locations 210
  MNEs and local systems, Italy 241
pharmaceuticals
  development costs 78–9
  expenditure 30–31
  investment 32–3
  marketing 33
  pricing 37, 51, 52, 77–97
pharmacies, and drug prices 79–80, 84, 94
physicians, ambulatory 28–9
Pierson, P. 60–61
Polanyi, K. 71
Porter, M. 188
pre-competitive approach to collaborative research 113
pricing of pharmaceuticals 37, 51, 52, 77–97
programming by integrated setting 68–70
property strategy
  Futuroscope 272
  Genopole 266–7
public
  expectations of health services 36
  opinion of British NHS 54–5
public funding
  health care 22–4
  university research 110
public–private partnerships, research 112–15

reciprocity and welfare systems 71
regional clusters, *see* clusters
regional impact analysis, research alliances 222–3
regional spillovers, California 226–9
regional technology poles, Europe 260–61; *see also* high-tech clusters
Reilly, R. 176
relational capital 132
research alliances 112–15, 220–31

Australia 116–19
biotech industry 138–9
Italy 119–22
University of California 226–9
research and development
biotechnologies 32–3, 136–40
France 261–76
investment 32–3, 172
medical equipment 32–3
research, university
industry partnerships, *see* research alliances
OECD 108–12
USA 212–13
return on assets methods (ROA) 173
Riccaboni, M. 191
risk, and intangible assets 186–7
Roberts, J. 130

San Diego, biotech clustering 227–8
Schiuma, G. 185
Schweitzer, S.O. 46, 47, 171
science parks 260
scientific research, *see* research
scientists, role in biotech companies 192–3
scorecard methods 173
Sen, A. 22
size, health sector 18–21
social regulation systems 70–73
specialist hospitals, benchmarking problems 160–62
spending, *see* health expenditure
spillovers
health industry 38
innovation 128–9
research alliances 220–31
Stephan, P.E. 190
Stewart, T.A. 172
Stiglitz, J. 56
Storey, D.J. 260
Storper, M. 187
strategic decision-making
and health systems 50–56
and industry promotion 4, 6
structural capital 131–2
Sugden, R. 50, 53
supply-related vertical interactions, clusters 225

tacit knowledge 183–4
Tambourin, P. 266
taxation, France, impact on companies 263–4
Taylor, R. 52
technology poles, regional 260–61; *see also* high-tech clusters
telemedicine 37
territorialization 187
Thomas, L. 141
Thomson, G. 132
trade, international, medical services 37
training, medical, and strategic decision-making 51–2

uncertainty and intangible assets 186–7
United Kingdom
biotech firm clusters 194
hospital costs benchmarking 150–65
intellectual property management 107–8
managed competition model 66
universities
and biotech firms 209–10, 213–16
California, economic impact of research 226–9
Ferrara, research activities 120–22
and Genopole 268–9
and knowledge management 103–22
research 108–22
collaboration with industry, *see* research alliances
university entrepreneurial approach to collaborative research 114
USA
biotech clusters 192–3
biotechnology industry 136–40
drug pricing 37, 78–96
health care spending 24–5
health insurance 29–30
health system 207
hospitals 27–8
pharmaceutical industry 135
research model 264
technology policy 258–9

venture capital
and biotech clusters, US 193
and intangible assets 187

vertical interactions, clusters 225
Viagra, controlled availability 51

Washington University, collaboration
with Monsanto 221–2
welfare regulation systems 70–73
Westhead, P. 260
Wolfram Cox, J. 153

Zeitlin, M. 50, 51
Zucker, L.G. 209